D1572849

FUNDAMENTALS OF PLANT GENETICS AND BREEDING

Fundamentals of Plant Genetics and Breeding

James R. Welsh
Colorado State University

John Wiley & Sons
New York • Chichester • Brisbane • Toronto

SB
123
.W4

Copyright ©1981, by John Wiley & Sons, Inc.

All rights reserved. Published simultaneously in Canada.

Reproduction or translation of any part of
this work beyond that permitted by Sections
107 and 108 of the 1976 United States Copyright
Act without the permission of the copyright
owner is unlawful. Requests for permission
or further information should be addressed to
the Permissions Department, John Wiley & Sons.

Library of Congress Cataloging in Publication Data:

Welsh, James R 1933-
Fundamentals of plant genetics and breeding.

Includes index.
1. Plant-breeding. 2. Plant genetics. I. Title.
SB123.W4 631.5′3 80-14638

ISBN 0-471-02862-2

Printed in the United States of America

10 9 8 7 6 5 4 3 2 1

To our children

Carrie, Mark, Michael,
Margie, Beverly and David

Preface

This book was written to provide the beginning plant breeding student with a basic set of principles. It can be used by those who only take an introductory course, but it can also serve as the foundation for more advanced study. Most chapters are purposefully brief. Plant breeding courses are traditionally taught in departments with active plant improvement programs and this text is designed so that the instructor can insert his or her own experiences and illustrations throughout the course. I hope this flexibility will show the individuality and creativity of plant breeding. Some of the more recent review articles that can be used to identify detailed readings are included and new science and technology of wide genetic transfers and tissue culture emerging in plant breeding and genetics are briefly discussed in Chapters 18 and 19.

The amount of basic genetics to be included was difficult to resolve. Selection of the topics on Mendelian genetics and related subjects in Chapters 2 to 5 was based on 15 years of experience in teaching plant breeding. Even those students who come into the introductory plant breeding course with a firm grasp of genetics can benefit by having a ready reference for quick review. This material should be especially helpful to students who have difficulty in understanding Mendelian principles. These chapters can be completely ignored or used in whatever manner best suited to the students' needs. The remainder of the book is designed to flow logically from one breeding principle to another and is not based on specific crops or pollination systems. For example, the pure line concepts are extended into inbred line improvement techniques, and the hybrid discussions include both cross- and self-pollinated crops. I hope this will make clear the understanding of principles without stressing traditional (and artificial) divisions of subject matter.

A few simple questions have been included at the end of some chapters. These are problem types that will aid the student in understanding material in the chapter.

Nomenclature and terminology are always troublesome and I have listed scientific names in a glossary at the end of the book to make the material more readable. The use of the terms "corn" versus "maize" and "variety" versus "cultivar" (cultivated variety) warrants a comment. Corn was chosen since it is a common name frequently used in U.S. literature. Maize is more botanically correct, however, and is used almost exclusively in European publications. The term variety was selected instead of cultivar because of its common usage, especially in the farming community. Cultivar is officially approved for scientific writing.

I gratefully acknowledge the many students and breeder colleagues who provided help and encouragement in the writing of this book. In particular, Dr. S. L. Ladd and Dr. D. D. Stuthman gave many valuable criticisms and suggestions after reviewing the entire manuscript. Dr. T. E. Haus evaluated several chapters and Dr. T. Tsuchiya supplied excellent chromosome photomicrographs. I recognize Colorado State University and my administrators Dr. W. F. Keim, Head of the Agronomy Department, and Dr. D. D. Johnson, Dean of the College of Agricultural Sciences, for making time available to complete this task. Finally, special recognition goes to my wife, Addie, who encouraged and worked with me through the entire manuscript preparation process.

Although many people provided information, any errors or omissions are exclusively my responsibility. This has been a pleasant endeavor—one that I hope will benefit plant breeding students.

JAMES R. WELSH

Contents

1

INTRODUCTION

People have always been engaged in some form of plant breeding by manipulating genetic systems to design a more desirable product. The definitions of improved characteristics are dictated by the user and lead to a wide potential group of breeding program goals. A producer of dryland grain crops, such as wheat or barley, and a gardener interested only in individual plants produced for aesthetic values will have different requirements and the breeder must respond to them. The array of challenges in plant breeding is extensive, and the number of alternative objectives necessitates establishing goal priorities based on careful study of total production and utilization. Efficient management of financial resources—a limiting factor in any program—is essential for breeding effectiveness.

As a profession, plant breeding offers many opportunities. Breeding programs can range from the work of the backyard hobbyist interested in the personal challenge of plant improvement to multimillion dollar programs designed to generate a financial advantage for a corporation. In competitive agricultural production, economic incentive is the underlying basis for plant breeding. Unproductive plant breeding programs that do not result in improved varieties or hybrids are often short-lived. Conversely, the hobbyist may be interested in developing plants for personal pleasure without particular regard to potential financial opportunities.

The professional breeder in a public or private program is a highly skilled and well-trained scientist. Successful breeders combine educational skills and training with intuition and a certain amount of luck—they are basically gamblers who attempt to improve the odds of a favorable outcome through a knowledge of genetics and breeding manipulations. They must know fundamentals of plant reproduction, function, and culture, and be able to use proper experimental techniques to insure a reasonable probability of making correct breeding decisions. Subject areas include genetics, agronomy, horticulture, botany, plant pathology, plant physiology and biochemistry, statistics, computer science, and economics. Even this partial list indicates that plant breeders cannot be expert in all pertinent subjects; they must recognize their limitations as well as strengths. Breeders often accept responsibility for intergrating and utilizing skills of other professionals in cooperative research efforts. Because of the potential benefit to society from genetic improvement of plant species, scientists in related fields are usually willing to cooperate in breeding and genetic activities.

Mechanical skills and imagination are required for plant breeders to develop equipment and techniques that achieve unique program objectives. The progressive, innovative breeder continually seeks maximum genetic gain with minimum resource input.

By improving food supplies plant breeding has played a major role in helping to reduce massive hunger and starvation on a worldwide basis. The Nobel Peace Prize awarded to Dr. N. E. Borlaug in 1970 is a testimony to the importance of plant breeding to human welfare. Dr. Borlaug's intense efforts in developing improved wheat varieties have led to extensive production increases in many calorie-deficient areas of the world. However, breeding efforts must be coupled with production and marketing improvements and adequate population control, if world food needs are to be properly confronted. As world population continues to increase, food production technology—including plant breeding—will struggle to keep pace with demand.

The advisability of genetically improving yields in crops where apparent surpluses exist is frequently debated. In the United States, for example, "overproduction" of many grain crops has resulted in depressed prices and the need for government subsidies to maintain adequate farm income. On the other hand, estimates indicate that a significant proportion of the world population experiences some form of hunger or malnutrition every day. In the world marketplace the fine balance between adequate and inadequate supplies of food was demonstrated in 1973. Dramatic price increases followed massive Russian government grain purchases necessitated by weather-related production shortfalls.

Production efficiency is the result of increased output per unit of input. Improved varieties contain built-in production stability factors provided at a minimum financial expenditure for both the producer and the consumer, and aid in insuring a steady food, feed, and fiber supply. Our total environment, including the plant species around us, has a direct effect on our psychological well-being and happiness. In ornamental species, the genetic manipulation of plant characteristics that provide pleasure and enjoyment is an important function of plant breeders.

These brief considerations emphasize that from humanitarian or economic standpoints the breeder has every obligation to continue genetic plant improvement. Plant breeding is the manipulation of a biological system that requires many generations to achieve results; it is also a dynamic, exciting, and challenging profession operating under continually changing conditions. The remainder of this book is devoted to principles and practices related to the art and science of plant breeding.

2

MENDELIAN GENETICS

Today's plant breeding utilizes for its basic foundation the genetic principles initiated by the classic investigations of Gregor Mendel. We have advanced significantly in our scientific knowledge of inheritance since his work, but we are still primarily using refinements of Mendelian principles. Mendel, an Austrian monk, was well trained in scientific investigations. His work, reported in 1865 (1), employed simple but effective approaches to the solution of inheritance problems that had troubled scientists for many years. It is important to understand these principles when addressing plant breeding problems and so this chapter is devoted to a detailed description of Mendel's investigations and conclusions. The sound techniques and logic he employed allow his work to stand as a model of methodology in the design of all scientific investigations.

Mendel initially studied a number of different plant species, but two important properties prompted him to choose the garden pea for his major work. First, he recognized several different traits that had repeatable variation between parents. For example, seed shape could be either round or wrinkled, and each variation of the seed-shaped trait was transmitted from generation to generation on a highly repeatable basis. The variations were easily recognizable and could be stabilized into true breeding forms or lines. Secondly, the plants had a flowering mechanism that provided protection, or could be easily protected, against the influence of all foreign pollen. He required that the offspring from a single plant were the result of the union of a male and female from that same plant or that they came from specific parental combinations under his control. The garden pea met the requirements for well-defined traits and a closed mating system.

The object of Mendel's experiment was to deduce the rules by which observable inherited variations in certain plant characters were passed from generation to generation. The term "character" means different plant traits or forms, and in his study included seed shape, seed color, flower color, pod shape, pod color, flower position, and stem length. His experiment initially resolved itself into as many separate experiments as there were different characters in which he was interested. Other investigators who had conducted earlier inheritance studies attempted to look at the organisms in their entirety. This approach introduced so much complexity into the system that they were unable to identify specific patterns and mechanisms. Mendel employed the sound scientific principle of reducing a complex question to its component parts for study, and then bringing the parts together for the final conclusions.

Mendel established parents that were true breeding for differing expressions of each character. He did this by selecting specific expressions for several generations until there was no further variation in the progeny. After he had established the true breeding parental lines, he made crosses, or hybridizations, between parents that differed in expression of one character. This was done separately for each of the seven characters. The resulting seeds were harvested and were the beginning of the first generation of the cross (F_1). The seeds were planted and the resulting plants were the F_1 generation. Seeds harvested from these plants were the initiation of the F_2 generation. In some cases he carried his material through the F_3. In all generations he kept seeds from each plant separate and maintained accurate records. This allowed him to classify and count the number of individuals occurring in each group. With this simple statistical technique, patterns in terms of ratios began to emerge.

We now know that some chance or random deviations will generally occur in any data set, and we use modern statistical techniques to understand the meanings of the deviations. Mendel used no statistical tests because they had not been developed at that time, and could draw conclusions only from patterns in his data.

MONOHYBRID SEGREGATION

Table 2.1 provides data for each of the seven initial experiments. Mendel used the term "dominant" to describe the expression of the character that appeared in the F_1 generation and "recessive" for the expression that disappeared in the F_1 but reappeared in the F_2. In each of the seven cases the pattern of 3 dominant : 1 recessive in the F_2 was closely approximated.

He then studied the F_3 generation by evaluating offspring from each F_2 plant. F_1, F_2, and F_3 data for the two characters of seed shape and seed color are given in Table 2.2. The F_2 plants that were recessive in appearance produced only recessive individuals in the F_3. Of the F_2 plants with dominant expression, approximately one-third yielded all dominant progeny, while two-thirds produced progeny that again segregated in a 3 dominant : 1 recessive manner. Combined F_3 information for any population produced an F_2 ratio of 1 pure dominant : 2 mixed dominant-recessive : 1 pure recessive. The term "monohybrid" is now commonly used to refer to the cross between parents with different expressions of a character that will result in a 3 : 1 ratio or some variation. We now know that this ratio combination is controlled by one inheritance unit called a gene.

Table 2.1 F_1 and F_2 Monohybrid Phenotypic Data For The Seven Characters Studied by Mendel

Character	F_1	F_2
Seed shape	Round	5474 round : 1850 wrinkled
Seed endosperm color	Yellow	6022 yellow : 2001 green
Flower color	Red	705 red : 224 white
Pod shape	Inflated	882 inflated : 299 constricted
Pod color	Green	428 green : 152 yellow
Flower position	Axial	651 axial : 207 terminal
Stem length	Tall	787 tall : 277 short

Table 2.2. F_1, F_2, and F_3 Dihybrid Data For Seed Shape and Seed Color

F_1	F_2	F_3
Round yellow	315 round yellow	138[a] : round yellow and green, wrinkled yellow and green
		60 : round yellow and wrinkled yellow
		65 : round yellow and green
		38 : round yellow
	101 wrinkled yellow	68 : wrinkled yellow and green
		28 : wrinkled yellow
	108 round green	67 : round and wrinkled green
		35 : round green
	32 wrinkled green	30 : wrinkled green

[a] Refers to the number of F_2 plants producing these expression combinations in their progeny. Totals are reduced slightly from the F_2 numbers because of some inviable seeds.

To label the variations of each character, Mendel developed the system of using capital and lowercase letters for the dominant and recessive expressions. For example, he assigned *A* to represent the dominant parent expression and *a* to the recessive parent. He then reasoned that the F_1 must be a combination of the potential expressions from each parent and should be designated *Aa.* The same reasoning was applied to the F_2 generation where one-half were true breeding *A* or *a.* The half that produced both kinds of offspring in the F_3 were labeled *Aa.* In many respects Mendel's selection of symbols for his inheritance language was unfortunate since it did not take into account the numerous variations of a single gene that can occur in nature. The question of these variations, called multiple alleles, is considered in Chapter 4.

By studying the data for all characters, and recognizing the consistency among characters, Mendel developed the law of segregation dealing with the different expressions of a single character. This law is stated as follows:

> *It is now clear that the hybrids (F_1's) form seed having one or the other of the two differentiating characters, and of these one-half develop again the hybrid form, while the other half yields plants which remain constant and receive the dominant or recessive character (respectively) in equal numbers.*

Simply put, this means that each F_1 produces offspring that may show either the dominant or recessive variation of the character, and of the total F_2 population one-half will breed true and the other half will again segregate exactly as the F_1's did. Although he did not specifically demonstrate the concept, Mendel suggested that the units of inheritance must occur in pairs in either parent, whether they were true breeding or hybrid.

DIHYBRID SEGREGATION

Following the investigation of each character individually, Mendel evaluated two characters at a time. True breeding parents with differing expressions were crossed and the F_1, F_2, and F_3 generations were evaluated. In some cases a few F_2 seeds were not viable so population numbers in the F_3 were not always exact totals of the F_2's. An example of the data is given in Table 2.2. In the F_2 all character combinations very closely approached a ratio of 9 double dominant : 3 recessive, dominant : 3 dominant, recessive : 1 double recessive. Some F_2 individuals were true breeding and would produce offspring like themselves for either or both characters. Others were hybrid and would produce assorted offspring for either or both characters. Based on his F_3 data, he suggested that the F_2 frequency distributions had the general pattern of $1AABB + 1aabb + 1AAbb + 1aaBB + 2AABb + 2aaBb + 2Aabb + 2AaBB + 4AaBb$, and that this pattern resulted from the simple multiplication of $(1AA + 2Aa + 1aa)(1BB + 2Bb + 1bb)$. Using these findings, Mendel developed the law of independent assortment dealing with the different expressions of two or more characters considered simultaneously. He states:

The offspring of the hybrids in which several essentially different characters are combined exhibit the terms of a series of combinations in which the developmental series for each pair of different characters are united. It is demonstrated at the same time that the relation of each pair of different characters in hybrid union is independent of the other differences in the two original parent stocks.

This simply means that the inherited differences in one characteristic are not influenced by the inherited differences in another character. Each event has an assigned mathematical probability with no one event being favored over, or influenced by, any other. The term "dihybrid" describes those inheritance cases where a 9 : 3 : 3 : 1 F_2 ratio, or a variation of this ratio, is produced. The investigation of three characters produced results that agreed with the inclusion of a third variable in the multiplication formula.

BACKCROSSING AND RECIPROCAL CROSSING

Mendel used two additional crossing programs to verify his results regarding the mechanics of inheritance systems. Data are presented in Table 2.3. Backcrossing (the crossing of an F_1 to one of the parents) demonstrated that the F_1 plants indeed contained an equal combination of potential expressions for each character. This was proven when all expressions of the character appeared with equal frequency in the progeny of the F_1 crossed with the recessive parent. For example, the F_1 with round yellow seed, when crossed to a true breeding double recessive wrinkled green parent, produced all four classes of progeny with approximately equal frequency. Mendel also demonstrated that this inheritance system operates on a predictable basis regardless of which parent is used as the male or female. In this reciprocal cross study he used the F_1 as the female for one set of data and as the male for the other. Results from the reciprocal crosses were almost identical.

RANDOMNESS OF EVENTS

In addition to developing the concept that the units of inheritance occur in pairs in each individual, Mendel suggested that

Table 2.3. Reciprocal Cross and Backcross Results of the F_1 Crossed with Round Yellow and Wrinkled Green True Breeding Parents

Cross		Progeny
♀[a] F_1	× ♂ round yellow parent	98 round yellow
	× ♂ wrinkled green parent	31 round yellow 26 round green 27 wrinkled yellow 26 wrinkled green
♂ F_1	× ♀ round yellow parent	94 round yellow
	× ♀ wrinkled green parent	24 round yellow 25 round green 22 wrinkled yellow 26 wrinkled green

[a] ♀ = female, ♂ = male.

each unit is passed separately through the male and female components of the reproductive mechanism (gametes) with each unit having an equal likelihood of being passed to the next generation. The mechanics of these concepts have since been expressed in systems such as the checkerboard (Punnett) square, the branching method, and the binomial expansion. To illustrate, we will consider the case of seed shape and incorporate the randomness and pairs concept into the system. Round seed is dominant and will be designated *A*. Wrinkled seed is recessive and will be identified as *a*. Two true breeding parents with contrasting expressions of the character are obtained by selection. If the units of inheritance occur in pairs, then each parent must have two identical units to be true breeding, in this case *AA* or *aa*. This condition in modern genetic language is termed homozygous. Each inheritance unit is transmitted with equal frequency to the offspring through the reproductive cells. Either parent passes each of its inheritance units

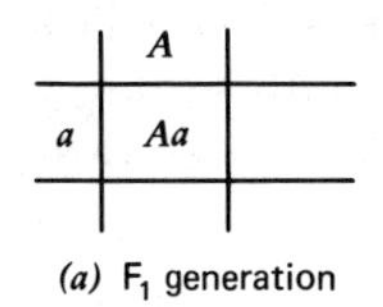

(a) F_1 generation

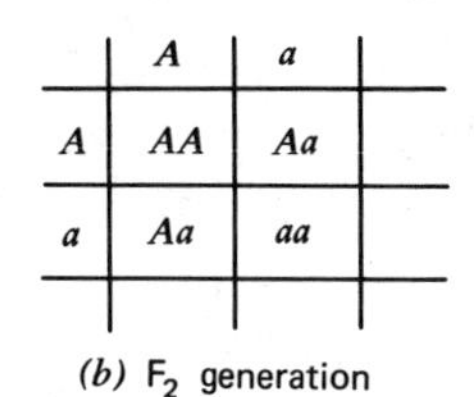

(b) F_2 generation

Figure 2.1. Segregation inheritance patterns from the cross *AA* × *aa* with reproductive cells (gametes) on the margins and progeny in the center.

with equal frequency but, since both are true breeding, there is only one option for each homozygous parent. This is illustrated in Figure 2.1*a* where the Punnett square has all possible contributions from one parent across the top and from the other parent down the side with the resulting offspring in the center. All hybrid or F_1 seeds will be *Aa* and round because, although they contain both *A* and *a*, the dominant will be expressed. Individuals possessing both gene forms are called heterozygous.

To obtain F_2 results, alternative gametes from the F_1 plants must be considered. The Punnett square for the F_2 population is diagramed in Figure 2.1*b*. Here each parental plant will contribute *A* or *a* with equal frequencies to the offspring. When the total number of progeny in each morphological class is determined, an F_2 ratio of 1*AA* : 2*Aa* : 1*aa*, or 3 round : 1 wrinkled is produced. This ratio can also be obtained mechanically by the expansion of the binomial $(A + a)^2$. These results, in agreement with those obtained by Mendel, are based on two assumptions. First, units of inheritance occur in pairs and, second, each unit is passed with equal frequency (randomly) to the next generation. The law of segregation is satisfied by these assumptions.

We will now consider independent assortment where characters segregate simultaneously and independently. As an example, the character of seed color will be combined with seed shape. Since yellow is dominant it will be designated *B,* while the recessive is green and termed *b.* Taken individually, the inheritance pattern of each character is exactly the same and monohybrid in nature. To combine the two characteristics, homozygous parents of round yellow, *AABB,* and wrinkled green, *aabb,* are crossed to produce the F_1. Each parent will transmit the inheritance units entirely at random but will be limited in alternatives because of homozygosity. This pattern is illustrated in Figure 2.2*a.* Each F_1 will have round yellow seed but will contain the alternative gene forms *A* and *a,* and *B* and *b* in equal frequency.

The fate of the F_2 population hinges on the production of gametes (eggs and pollen) by the F_1 individuals. If the control of seed shape and seed color is independent, and if the contrasting gene forms for each character are passed with equal frequency, then the inheritance unit combinations should occur in a ratio of 1*AB* : 1*Ab* : 1*aB* : 1*ab* in the reproductive cells. The validity of this assumption is checked by constructing a Punnett square to produce all F_2 progeny with their proper frequencies. The results are shown in Figure 2.2*b.*

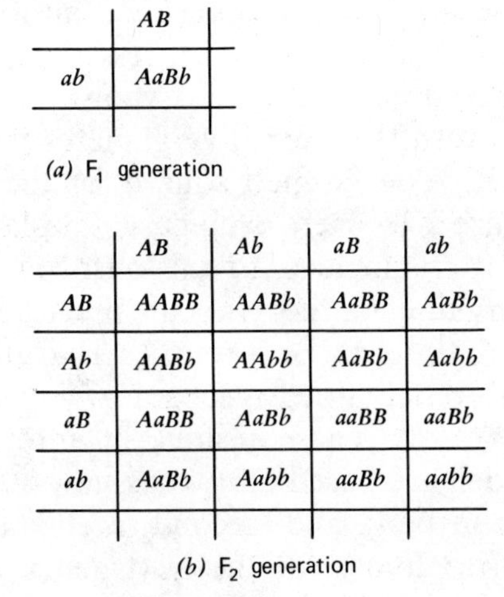

	AB
ab	*AaBb*

(*a*) F_1 generation

	AB	*Ab*	*aB*	*ab*
AB	*AABB*	*AABb*	*AaBB*	*AaBb*
Ab	*AABb*	*AAbb*	*AaBb*	*Aabb*
aB	*AaBB*	*AaBb*	*aaBB*	*aaBb*
ab	*AaBb*	*Aabb*	*aaBb*	*aabb*

(*b*) F_2 generation

Figure 2.2. Independent assortment inheritance from the cross *AABB* × *aabb* with reproductive cells (gametes) on the margins and progeny in the center.

By collecting identical terms in the F_2 we obtain the ratio of:

1 *AABB*	Round yellow
1 *AAbb*	Round green
1 *aaBB*	Wrinkled yellow
1 *aabb*	Wrinkled green
2 *AaBB*	Round yellow
2 *Aabb*	Round green
2 *AABb*	Round yellow
2 *aaBb*	Wrinkled yellow
4 *AaBb*	Round yellow

These data can also be expressed in a ratio of 9 round yellow : 3 round green : 3 wrinkled yellow : 1 wrinkled green and agree almost exactly with those obtained by Mendel. Thus, the inheritance mechanism proposed for single characters can validly be extended to include several characters at the same time without destroying its validity—providing the assumptions of independence, randomness, and pairs of units are not violated.

The ratios produced by the Punnett square can also be generated by the binomial expansion of $(A + a)^2 (B + b)^2$. Note that this expansion also includes the properties of independence, randomness, and pairs of units. Another common and quite easy method of constructing F_2 population ratios is called the branching system. It is illustrated in Figure 2.3. Again the three basic assumptions are included.

Our working vocabulary will now be expanded to include the term "genotype," which is the genetic makeup and "phenotype," which is the physical appearance of the individual. Genotypic ratios and phenotypic ratios can be generated in very similar manners. For example, in a dihybrid F_2 population, phenotypic combinations with their frequencies can be obtained by the binomial expansion of (3 round + 1 wrinkled) (3 yellow + 1 green) or by the branching method shown in Figure 2.3. These systems work nicely for small numbers of characters. Beyond about three, however, they become very cumbersome and difficult to handle accurately.

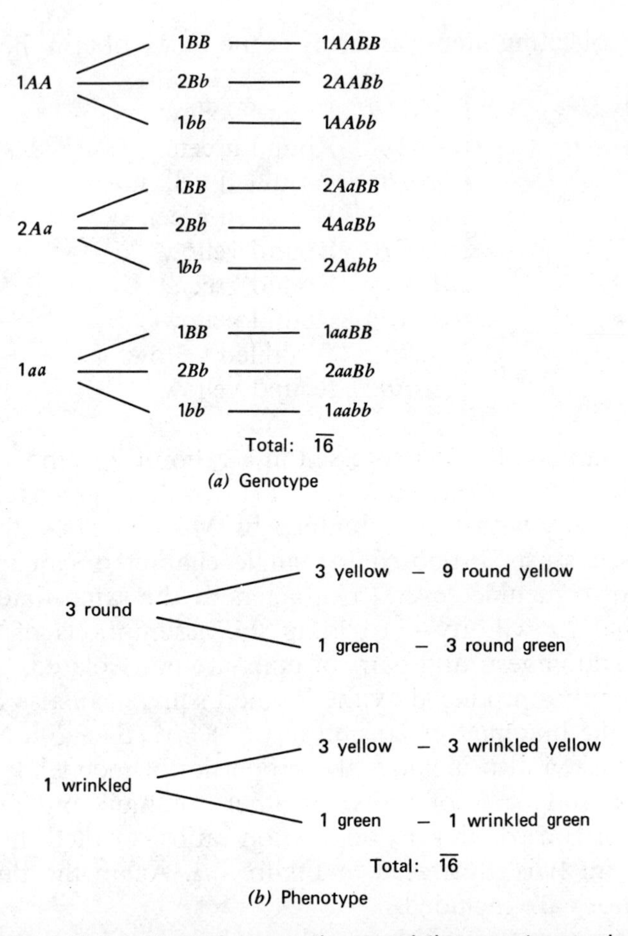

Figure 2.3. Branching method of segregating population genotype and phenotype construction. In this example a dihybrid population is used.

Some shortcut methods have been proposed by Mendel and others to describe F_2 populations. If n = the number of characters, then 2^n = the number of true breeding individuals in the F_2, 3^n = the number of different genotypic combinations in the F_2, and 4^n = the total number of individuals in the F_2 ratio. We can illustrate the use of these formulas by considering two characters ($n = 2$) independently assorting in the F_2. The number of true breeding individuals is 2^2 or 4. There are 3^2 or 9 different genotypic combinations. The total number of individuals in the F_2 ratio is 4^2 or 16. These results can be checked against those in Figure 2.3. The proportion of any term in the F_2 population can be easily determined. All homozygous genotypes have a frequency value of 1. The frequency of each genotype with heterozygous loci is determined by multiplying the number of heterozygous loci by 2. Consider the case of seven segregating characters. If we wish to know the frequency of *AABbCCDdeeffgg* individuals, two steps are necessary. In this case the term has a value of 4, since there are two heterozygous combinations. The number of individuals in the total F_2 population is 4^7 or 16,384. Thus the frequency of this genotype is 4/16,384 or 1/4096. By careful manipulation of the various formulas, complex genetic situations can be easily resolved into their component parts.

SUMMARY AND COMMENTS

Through an elegant set of experiments, Mendel was able to accurately describe inheritance mechanisms based on the assumptions of paired units and random transmission of the units from parent to progeny. While the system has been extensively refined through detailed study, the laws of segregation and independent assortment are as valid today as they were in 1865.

The beautifully sound experimental techniques and logic that Mendel employed can serve as a model for us all to follow in plant breeding and research programs. I often lose sight of the original purpose in an experiment because of the many interesting questions raised with each new set of data. It is helpful to remember that Mendel was successful in producing his great contributions by carefully working out the more simple details first and then combining the component parts to describe the complete, complex picture.

REFERENCES

1. Mendel, G. J. 1865. *Versuche uber Pflanzen-Hybriden.* Verh. L. Naturforrschenden Vereine in Brunn. Available in the original German as Vol. 42 (1), 1951 *Journal of Heredity.* English translation available from the Harvard University Press, Cambridge, 1948, and in James A. Peters (ed.), *Classic Papers in Genetics,* Prentice-Hall, Inc., Englewood Cliffs, N.J., 1962.

QUESTIONS

1. Consult Table 2.1. Use the characters of seed shape (*A* or *a*), seed color (*B* or *b*), and flower color (*C* or *c*). Construct the F_2 population composition for both genotype and phenotype using the Punnett square, the branching method, and the binomial expansion. What are the frequencies of the following genotypes?
Aa BB cc **Answer:** 1/32
aa Bb Cc **Answer:** 1/16
aa bb cc **Answer:** 1/64
What are the frequencies of the following phenotypes?
Round, yellow, red **Answer:** 27/64
Wrinkled, yellow, red **Answer:** 9/64
Wrinkled, green, white **Answer:** 1/64

2. An F_2 population resulting from the intermating of several round green red F_1's produces a ratio of 97 wrinkled green white : 870 round green red : 310 round green white : 294 wrinkled green red. What were the genotypes and phenotypes of the true breeding parents crossed to produce the F_1? **Answer:** There are two possibilities. *AA bb CC* (round green red) × *aa bb cc* (wrinkled green white), or *AA bb cc* (round green white) × *aa bb CC* (wrinkled green red).

3. Assume five characters, each controlled by a single gene pair independently assorting. Calculate the following properties of an F_2 population:
Total number of individuals: **Answer:** 1024
Proportion of true breeding individuals: **Answer:** 32/1024
Number of different genetic combinations: **Answer:** 243

4. What proportion of the F_2 population in Question 3 is in each of the following genotypes?
AA bb CC Dd Ee **Answer:** 1/256
AA bb cc DD ee **Answer:** 1/1024

3

CHROMOSOMES

Darkly staining structures called chromosomes within the nucleus of the cell were first described in the 1870s. Because of their accurate duplication ability and constancy of numbers, it became evident that a relationship existed between this highly regular mechanism and Mendel's observations on inheritance. Early workers showed that chromosomes and genes moved together. They observed that mitotic (duplication division) chromosome number remained constant but that chromosomes occurred in pairs and the number was halved during meiosis (reduction division) in the formation of reproductive cells called gametes. When the gametes combined to form the embryo, the chromosomes returned to the original number.

Chromosomes have been the subject of intense investigations by cytologists and cytogeneticists. With the improvement of microscopic techniques and resolution, chromosomes have been described in great detail. This chapter is devoted to chromosome movement and function.

CHROMOSOME COMPOSITION AND FUNCTION

Chromosome composition and function are important considerations in plant breeding. Chromosomes are double helix spirals composed of deoxyribonucleic acid (DNA), which is made up of organic base compounds of adenine, thymine, cytosine, and guanine attached to sugar molecules and linked together with phosphorus. The organic bases form the letters of the genetic language of life. Genes are defined in terms of the sequence of bases in a

DNA segment. The double helix structure of complementary strands has the capacity to reproduce itself very accurately by using the strands as templates. Thus, DNA can be duplicated repeatedly, and each duplication can be passed from parent to offspring. The duplicating property of DNA results in an almost error-free information transfer from one generation to the next.

Functionally, DNA provides information from the genetic code to drive the physiological mechanisms of the organism. This information is obtained by "reading" the coded sequence from partially opened DNA strands and translating it through another system called RNA (ribonucleic acid) to be expressed as the formation of a particular enzyme protein having a specific function in cell operation. DNA information, in combination with environment, results in the final phenotypic expression of each individual.

MITOSIS

Mitosis is a duplication division where cells reproduce themselves with the same chromosome number in the daughter cell as in the parent. Figure 3.1 illustrates mitosis in barley that has a somatic (nonreproductive) cell chromosome number of 14. The somatic chromosome number is written $2\underline{n} = 14$. While cell division is divided into four major stages for description purposes, it is really a continuous process with one stage integrated into the next. The stages are defined as follows.

Interphase The chromosomes are indistinguishable from one another and the nucleus stains as one body. This is the most active chromosome stage in the function of physiological mechanisms. During this stage, the gene information is read and translated to the biochemical mechanisms of the organism. The chromosomes are surrounded by a nuclear membrane that separates the nucleus from the remaining contents of the cell (cytoplasm).

Prophase The chromosomes prepare for division by shortening and thickening (Fig. 3.1*a,b,c*). Chromatids (longitudinal half chromosomes resulting from duplication) can be detected. Nucleoli (circular darkly staining bodies) can be seen in this stage. They are associated with specific points on certain chromosomes and

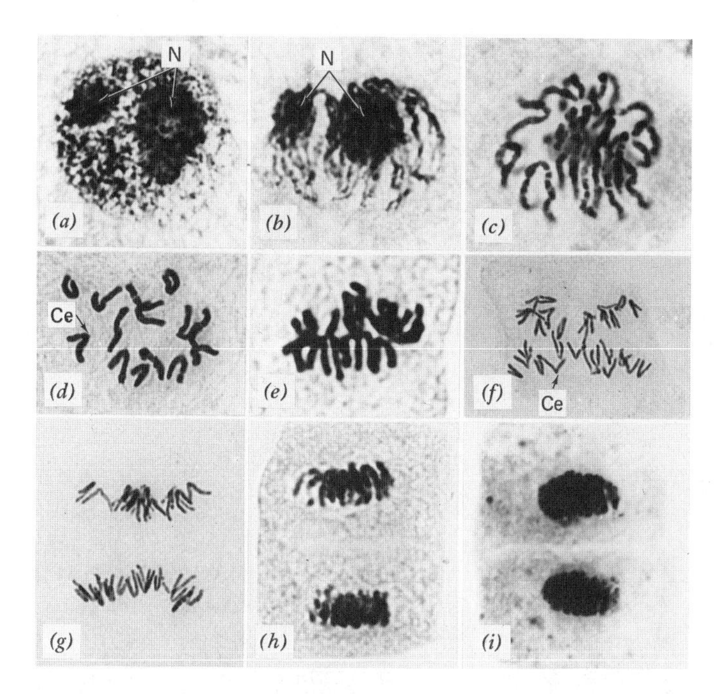

Figure 3.1. Stages of mitosis in barley (2n = 14). (*a*) Early prophase. (*b*) Mid-prophase. (*c*) Late prophase. (*d*) Early metaphase. (*e*) Late metaphase. (*f*) Early anaphase. (*g*) Late anaphase. The chromosomes are moved by centromere attachment to the spindle. (*h*) Early telophase. (*i*) Late telophase. A cell wall is forming between the new cells and interphase will occur. N-Nucleolus, Ce-Centromere. (Courtesy T. Tsuchiya, Colorado State Univ.)

are important in chromosome organization and cell division mechanisms.

Metaphase The chromosomes arrange themselves in a random manner on the equitorial plate or central plane of the cell (Fig. 3.1*d,e*). By the initiation of this stage the nuclear membrane and nucleolus have disappeared. The centromere, a region vital to chromosome movement, becomes attached to the spindle fiber that is responsible for directional chromosome movement during division.

Anaphase The centromere splits lengthwise in the chromosome and the chromatids begin to move on the spindle fiber toward the poles (Fig. 3.1*f,g*), with the centromere leading the movement. The individual chromatids are now considered new chromosomes.

Telophase The new chromosomes have completed their movement toward the poles and begin to disperse inside the nuclear membrane (Fig. 3.1*h,i*). As this stage proceeds a new cell wall is initiated between the two new nuclei.

Interphase This is the completion of mitosis and two identical daughter cells have been produced from one original cell.

Mitosis contains two important features. First, the chromosomes (and thus the genes) reproduce and divide so that the daughter cells contain exactly the same genetic information as the parent cell. As the division progresses, the centromeres split so there are as many in the daughter cells as in the original cells. Mitosis plays a key role in such biological processes as growth, replacement of damaged cells, and tissue repair.

MEIOSIS

A more interesting cell process from the genetic standpoint is meiosis or reduction division that produces gametes. Gametes formed by meiosis can combine to produce a new embryo containing equal amounts of genetic information from each parent.

The entire meiotic process involves two separate divisions called meiosis I and meiosis II. Meiosis I is the true reduction division where chromosome numbers are halved. Meiosis II is a mitotic or duplication division. In both divisions, the stages are identified in the same general manner as in mitosis. However, prophase in the first meiotic division (prophase I) has five subdivisions since it contains several unique and important events. Steps in the meiotic process are illustrated in Figure 3.2. Barley, with a chromosome number of $2\underline{n} = 14$, is again used. The following description places emphasis on important features.

Interphase Interphase in meiosis is exactly the same as interphase in mitosis. The chromosomes are in a resting stage with respect to division but are active metabolically.

Prophase I As in mitotic prophase the chromosomes prepare for division by constricting and moving to the central plane of the cell. However, in meiotic prophase the special feature of pairing comes into play. The suffix on the name of several prophase I stages can be either "tene" or "nema" and both have been used in texts. Our description uses "tene."

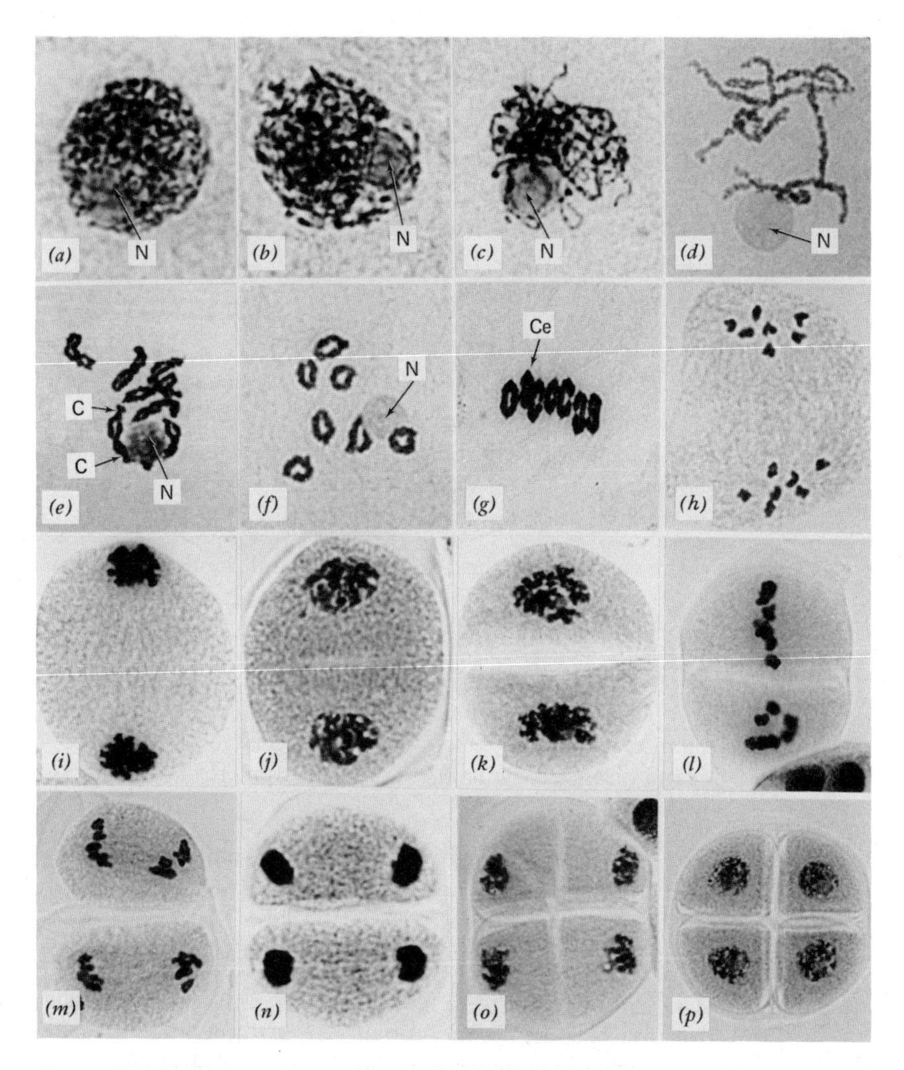

Figure 3.2. Stages in meiosis of barley (2n = 14). (*a–f*) Prophase I. (*a*) Leptotene. (*b*) Zygotene. (*c*) Early pachytene. (*d*) Late pachytene. (*e*) Diplotene. (*f*) Diakinesis. (*g*) Metaphase I. (*h*) Anaphase I. (*i*) Telophase I. (*j*) Interphase I. (*k*) Prophase II. (*l*) Metaphase II. (*m*) Anaphase II. (*n*) Early telophase II. (*o*) Late telophase II. (*p*) Interphase II. In anaphase I (*h*) reduction division occurs. N-Nucleolus, C-Chiasma, Ce-Centromere. (Courtesy T. Tsuchiya, Colorado State Univ.)

Leptotene The chromosomes begin to shorten and thicken slightly and become visible microscopically (Fig. 3.2*a*).

Zygotene The two identical or homologous chromosomes of each pair begin to associate (synapse) at various places along their length. The nucleolus becomes easily visible and is associated with specific regions on certain chromosome pairs (Fig. 3.2*b*).

Pachytene Synapsis is nearly complete along the length of the chromosomes and they continue to contract and thicken (Fig. 3.2*c,d*).

Diplotene The two chromatids of each chromosome become visible. The paired chromosomes begin to separate from one another. Points of very close association called chiasma (plural: chiasmata) are visible (Fig. 3.2*e*).

Diakinesis The centromeres push farther apart and the doubled chromosomes take on the appearance of rings. The chromosomes continue to contract, and the nucleolus begins to disappear (Fig. 3.2*f*).

Metaphase I The chromosomes move to the center of the cell, arrange themselves on the metaphase plate, and the nuclear membrane disappears (Fig. 3.2*g*). The centromeres attach to the spindle fibers.

Anaphase I The chromosomes begin to move toward the poles. In this stage the centromeres do not split and the chromatids associated with each centromere move as a unit. The actual reduction of chromosome numbers occurs in this stage (Fig. 3.2*h*).

Telophase I The chromosomes form into new nuclei and are surrounded by a nuclear membrane. A cell wall is formed between the two nuclei. Each nucleus at this stage is now $1\underline{n}$ rather than $2\underline{n}$ (Fig. 3.2*i*).

Interphase I The duration of this stage depends on the species. Two $1\underline{n}$ cells are now present. The meiotic process is ready to move into the second division (Fig. 3.2*j*).

Prophase II The chromosomes again shorten and thicken in preparation for the second division. Nucleoli are present and the nuclear membrane disappears (Fig. 3.2*k*).

Metaphase II The chromosomes arrange themselves on the metaphase plate (Fig. 3.2*l*).

Anaphase II The centromeres split lengthwise in the chromosomes and begin their movement to the poles, with one chromatid attached to each centromere half (Fig. 3.2*m*). These are now considered new chromosomes.

Telophase II The chromosomes are separated into new nuclei with the accompanying nuclear membrane enclosure (Fig. 3.2*n,o*). Cell walls are formed and the four reproductive cells are arranged in a formation called a quartet.

Interphase II The cells (Fig. 3.2*p*) are now ready for the final development into eggs or pollen grains. This will be described in Chapter 5 on plant reproduction.

Because of the undivided centromere and associated chromatids, lack of clearness often exists regarding true chromosome numbers in meiotic cells at various stages of division. A simple way to avoid confusion is by counting the centromeres in the cell and equating this to the chromosome number. Note that the $2\underline{n}$ number of centromeres is present in all stages of division in mitosis. In meiosis, $2\underline{n}$ centromeres are retained until anaphase I where they move intact to the resulting daughter cells. In meiosis II, a mitotic division, the centromeres split lengthwise and each new centromere with its respective chromatids (a new chromosome) is passed on to the gamete.

SEGREGATION AND INDEPENDENT ASSORTMENT

A significant aspect of meiosis is chromosome number reduction from $2\underline{n}$ to $\underline{n}$. This reduction system results in one chromosome of each pair being present in each reproductive cell at the end of meiosis II and allows for the return to the $2\underline{n}$ number when gametes combine to form a new zygote. The reduction system based on pairing becomes especially important since genes controlling inherited variation will be transmitted via the chromosomes.

The randomness of segregation and independent assortment

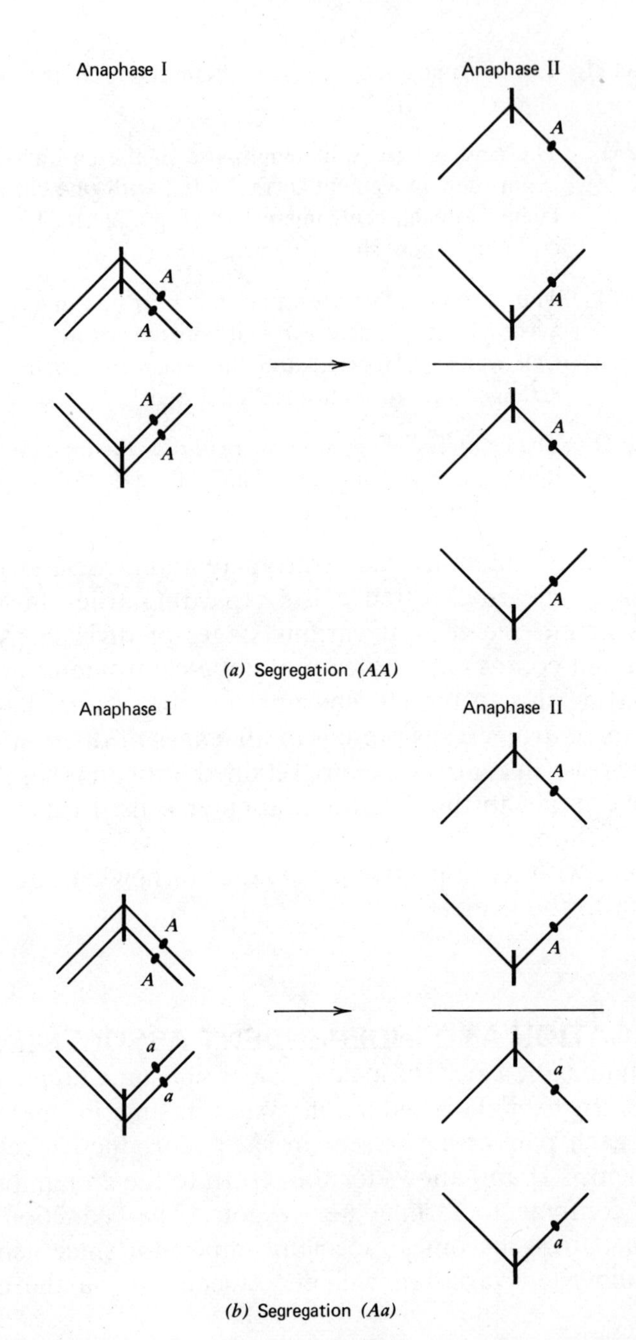

Figure 3.3. Anaphase I and anaphase II illustrating segregation and independent assortment.

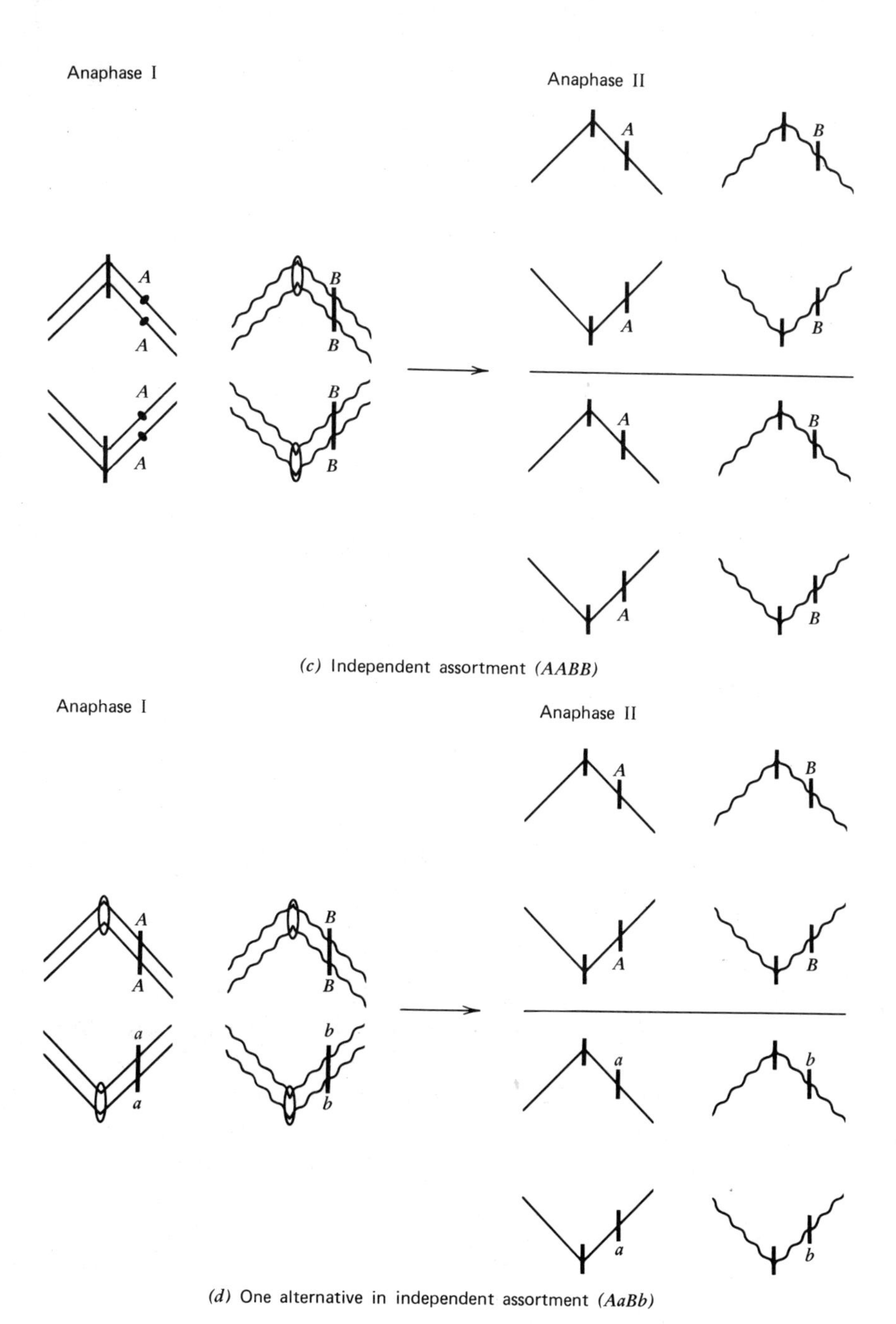

(c) Independent assortment *(AABB)*

(d) One alternative in independent assortment *(AaBb)*

Figure 3.3. (*Continued*)

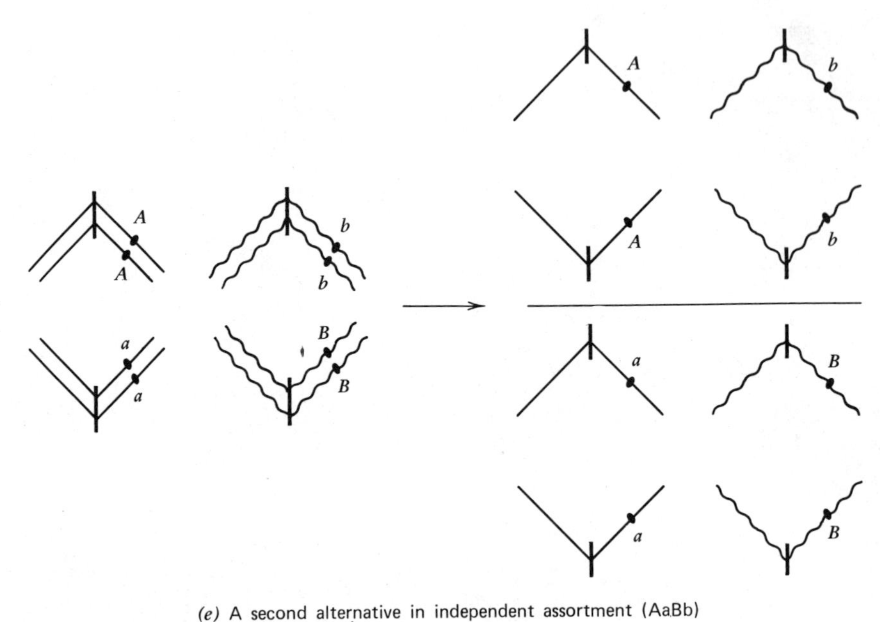

(*e*) A second alternative in independent assortment (AaBb)

Figure 3.3. (*Continued*)

is based on the direction, dictated purely by chance, in which each chromosome moves to the pole in anaphase I. In Figure 3.3 various alleles are placed at one or more loci and anaphase I and anaphase II are illustrated.

In the case of a characteristic controlled by one locus, a homozygous dominant individual (*AA*) will produce gametes from meiosis as shown in Figure 3.3*a*. Note that all gametes have the same genetic makeup for the *A* allele. An individual homozygous for the recessive allele (*aa*) at the same locus will produce gametes that are all recessive. These would be the equivalent of Mendel's true breeding parents.

The results of meiosis in a heterozygous individual (*Aa*) are shown in Figure 3.3*b*. An equal number of gametes will receive the *A* or the *a* allele. When the gametes are combined in all possible combinations (at random) the 1*AA* : 2*Aa* : 1*aa* ratio results.

Mendel's requirement of equal frequency gamete production is satisfied by meiosis.

Consider the dihybrid situation where each of the two different characters is controlled by a single locus, with dominance at each locus, and with each locus located on a different chromosome pair. Figure 3.3*c* shows the results of meiosis in an individual homozygous dominant at both loci. Chromosome reduction from 2n to n occurs in a random manner between members of each pair. Only *AB* gametes are produced, however, because of the limitations imposed by homozygosity. The limitation statement is true for the homozygous individuals of *aabb, AAbb,* or *aaBB* because they can produce only one type of gamete specific to their genotype.

For an individual heterozygous at both loci (*AaBb*) several possibilities exist. Figure 3.3*d* shows one of two alternatives. Here the chromosome carrying the *A* allele *happens by chance* to move in the same direction as the chromosome carrying the *B* allele, while the chromosomes carrying the recessive alleles move together in the opposite direction. This particular event results in the formation of *AB* and *ab* gametes with equal frequency. However, the direction of movement is a random situation and it is just as likely that the chromosome of one pair carrying the dominant allele would have moved with the chromosome of the other pair carrying the recessive allele (Fig. 3.3*e*), which results in equal numbers of *Ab* and *aB* gametes. Since these movement patterns occur equally often, *AB, Ab, aB,* and *ab* are produced with identical frequencies.

It is sometimes helpful to consider the meiotic cell as a sphere with poles at the top and the bottom and metaphase I chromosomes "floating" in the central plane. Each chromosome pair can then separate independently of any other pair. Because of random chromosomal movement, Mendel's law of independent assortment operates successfully. Remember that equal frequency distributions are observed only through the measurement of several meiotic divisions and their resulting gametes, not through a single division. Thus, when Mendel recorded F_2 data, he sampled large enough populations to observe the results of equal frequency independent assortment gametes.

LINKAGE

Peas have a $2\underline{n} = 14$ chromosome number. Considering the number of characters that make up a complex organism like the pea plant, it becomes apparent that many genes must be located on one chromosome. Genes on the same chromosome are linked and tend to be inherited in blocks or groups rather than independently. Independent assortment ratios of various progeny classes no longer occur since the assumption of randomness has been negated. To illustrate the effects of linkage, we will assume that resistance to a disease is controlled by a single locus with dominance for resistance (*R*) and recessive susceptible (*r*), and another locus governs seed color with yellow dominant (*Y*) to green (*y*). Anaphase I and anaphase II in Figure 3.4 will demonstrate meiosis with linkage.

If a parent is homozygous for resistance and yellow seed (*RRYY*) it will produce only *RY* gametes as illustrated in Figure 3.4*a*. Likewise, a parent homozygous for susceptibility and green seed (*rryy*) would produce only *ry* gametes. When these two parents are mated, the F_1 will be resistant with yellow seed (*RrYy*). Up to this point the system is indistinguishable from Mendel's independent assortment principles.

Several possibilities occur in F_1 gamete production. We know that exchange of chromatid material can occur between chromosomes of a pair during meiosis. This phenomenon, called a "crossover" is highly correlated with the chiasmata frequency in prophase I. A crossover can change the allelic relationship between loci on the same chromosome. Figure 3.4*b* shows the potential gametes produced if *no* crossover occurs between the two loci. In this case, only *RY* and *ry* gametes are produced. They are called parental type gametes because they are identical to those produced by the original parents crossed to give the F_1. Figure 3.4*c* shows the results of a crossover between the *R* allele and the *y* allele on one chromatid of each chromosome. At the end of meiosis, four gametes are produced: *RY, Ry, rY,* and *ry*. *RY* and *ry* are parental gametes, while *Ry* and *rY* are crossover or recombination gametes because they result from a crossover in the region

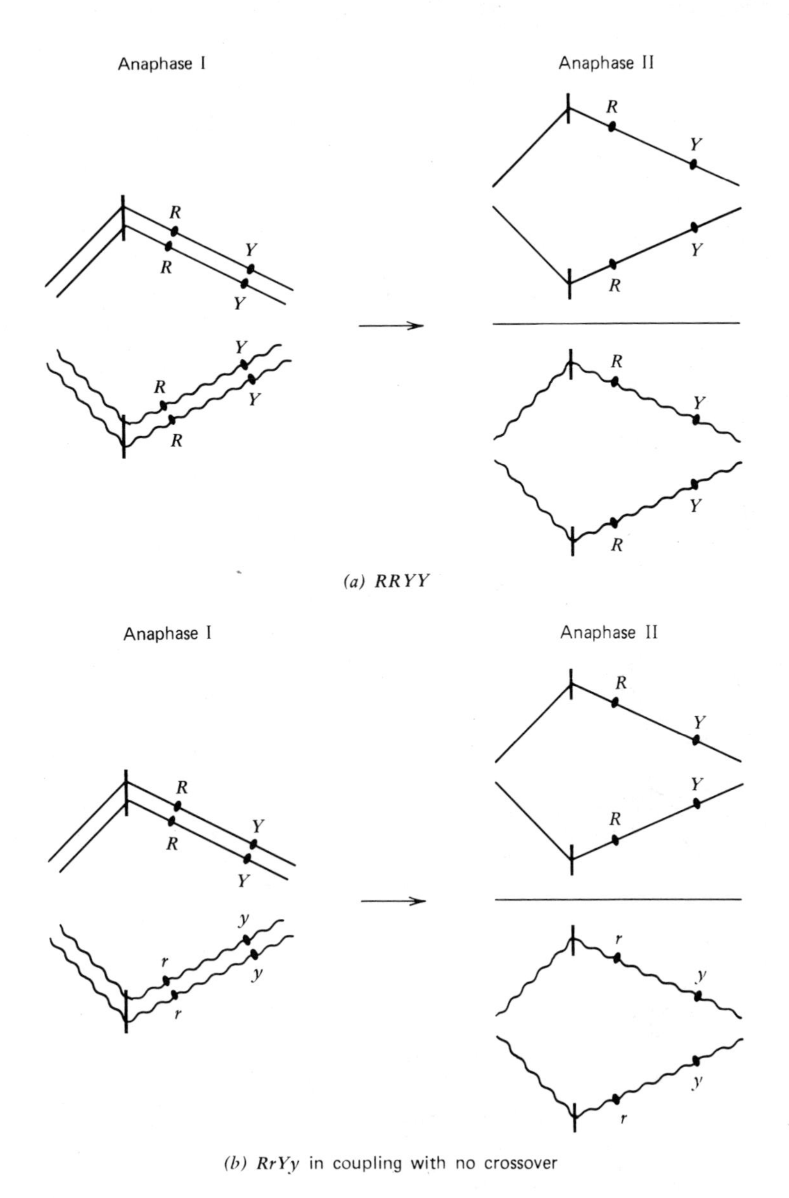

(*a*) *RRYY*

(*b*) *RrYy* in coupling with no crossover

Figure 3.4. Anaphase I and anaphase II illustrating linkage.

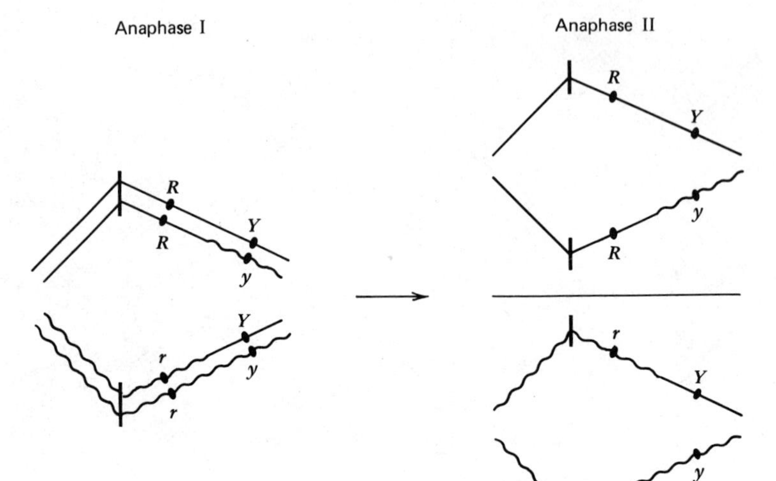

(c) RrYy in coupling with crossover

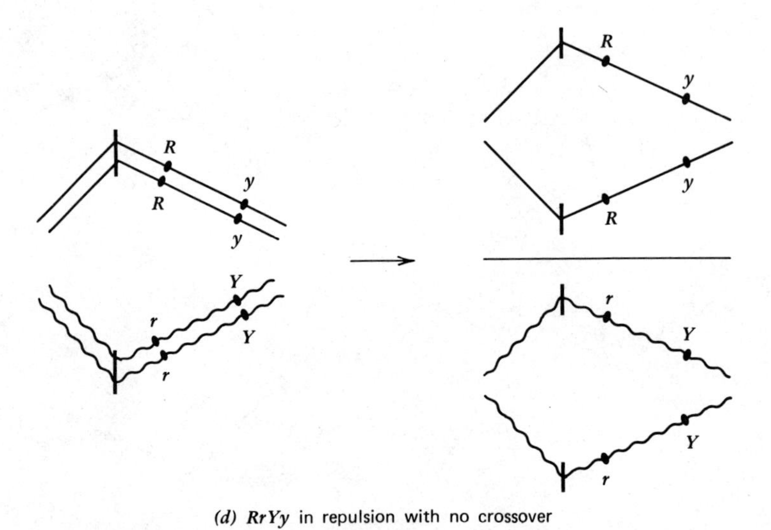

(d) RrYy in repulsion with no crossover

Figure 3.4. (*Continued*)

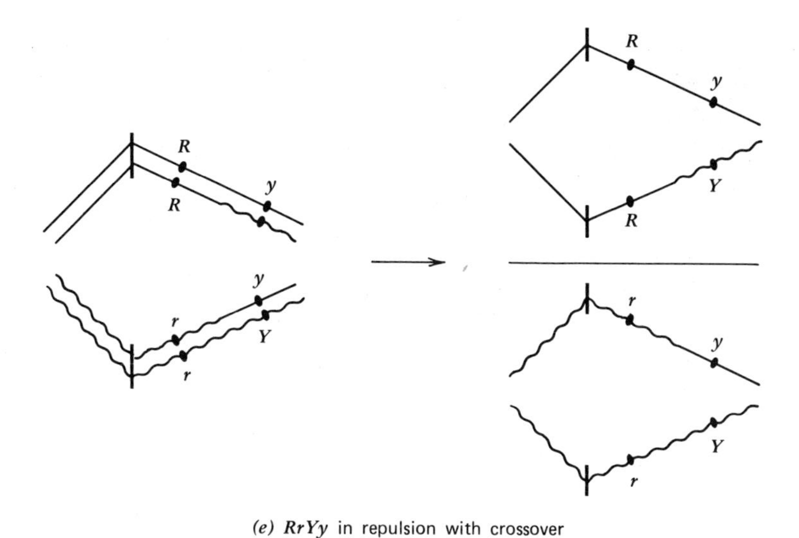

(e) $RrYy$ in repulsion with crossover

Figure 3.4. (*Continued*)

between the two loci. The probability of a crossover occurring in any region is primarily a function of the distance between the two loci, although other factors such as centromere proximity, distance to another crossover, and chromosome structural abnormalities can reduce crossover frequencies.

How do we know if two loci are linked or assorting independently? The answer lies in the frequency with which each gamete is produced. In the situation where the two loci are located on different chromosomes, the four gamete types of RY, Ry, rY, and ry are produced in equal frequencies. Equality of gamete production is the underlying assumption associated with Mendel's principle of independent assortment.

Now consider the linkage situation in Figure 3.4*c*. If no crossover occurs between the two loci, only RY and ry gametes will be produced resulting in exclusively parental types. A cell with a crossover between the loci gives a $1RY : 1Ry : 1rY : 1ry$ gamete ratio. Adding the results of both meiotic events together, a gamete ratio of $3RY : 3ry : 1rY : 1Ry$ is obtained. This ratio is also highly distorted from the 1 : 1 : 1 : 1 frequency of independent assortment. Depending on the distance between the two loci, the frequency of a crossover occurring between these two loci may vary from one in every cell to almost none in any cell.

We must now determine if gamete production is equal to or distorted from independent assortment. Two methods of measurement used to detect and estimate linkage are testcrossing and F_2 ratio analysis. In a testcross, the heterozygous F_1 (*RrYy*) is mated with a homozygous recessive (*rryy*). The progeny from this cross are then studied to determine the frequency of each gamete produced by the F_1. The testcross Punnett square is given in Figure 3.5. Each of the gametes from the F_1 has a distinct testcross progeny phenotype and can be identified immediately. If the F_1 is producing all gametes in equal frequency, then each of the four classes of testcross progeny will also occur with equal frequency. However, if the two loci are linked, parental gametes will appear with a greater frequency than crossover gametes and the ratio will be distorted in favor of the parental gamete types.

We know that some deviation from perfect ratio expression will always occur because of sampling error. The critical question is whether the observed deviations are sampling error or true deviations caused by linkage.

A simple statistical test called chi-square (χ^2) can be used to evaluate the amount of deviation from the expected equal frequency ratio in each class. The chi-square test can be used to test any ratio goodness of fit. In using this analysis on the testcross data, the hypothesis of independent assortment is checked and equal frequency classes are assumed. The calculated chi-square value is obtained by the formula $\chi^2 = \sum[(O-C)^2/C]$ where O is

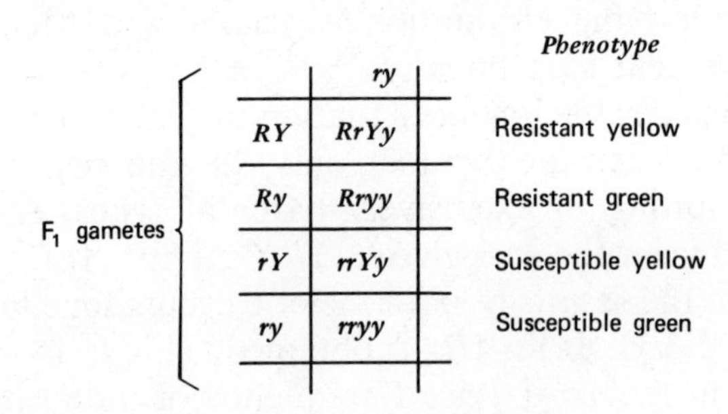

Figure 3.5. A testcross Punnett square from a cross of *RrYy* × *rryy*.

Table 3.1. Table of χ^2

n	P=.99	.95	.50	.10	.05	.01
1	.000157	.00393	.455	2.706	3.841	6.635
2	.0201	.103	1.386	4.605	5.991	9.210
3	.115	.352	2.366	6.251	7.815	11.345
4	.297	.711	3.357	7.779	9.488	13.277
5	.554	1.145	4.351	9.236	11.070	15.086
6	.872	1.635	5.348	10.645	12.592	16.812
7	1.239	2.167	6.346	12.017	14.067	18.475
8	1.646	2.733	7.344	13.362	15.507	20.090
9	2.088	3.325	8.343	14.684	16.919	21.666
10	2.558	3.940	9.342	15.987	18.307	23.209
20	8.260	10.851	19.337	28.412	31.410	37.566
30	4.953	18.493	29.336	40.256	43.773	50.892

n = degrees of freedom, P = probability.

Source: Reprinted with permission by Macmillan Publishing Co., Inc., from *Statistical Methods for Research Workers* by R. A. Fisher. Copyright © 1970, University of Adelaide.

the observed number or the real data and C is the calculated or expected frequency based on the hypothesis of independent assortment. In each class the deviation from the expected value is squared, divided by the expected value, and summed. We then compare this value with a preconstructed array of values (Table 3.1) that has been computed on chance deviation alone at several probability levels. If the deviations in the data are likely due to sampling error instead of a true difference in data from the expected values of independent assortment, the calculated chi-square will be equal to or smaller than the tabular values. If the computed values are larger than the tabular values, then the deviations are judged to be caused by something other than sampling error—in this case linkage. When using the tables to determine probabilities, the number of degrees of freedom is always one less than the number of classes or, in this case, three.

Table 3.2 gives an illustration of testcross data that have a very good fit to the independent assortment hypothesis. The total

Table 3.2. Sample Testcross Data With Independent Assortment or Wide Linkage

Phenotype	O	C	$(O-C)^2$	$\frac{(O-C)^2}{C}$
Resistant yellow	29	30	1	0.03
Susceptible green	32	30	4	0.13
Resistant green	28	30	4	0.13
Susceptible yellow	31	30	1	0.03
	120			0.32

calculated chi-square value of 0.32 has a 0.95 to 0.99 probability that the deviations result from chance alone. Table 3.3 gives data that fit very poorly to the hypothesis of independent assortment. The calculated chi-square value of 34.99 has less than a 0.01 probability of supporting the hypothesis of independent assortment that results in a 1 : 1 : 1 : 1 ratio.

A new hypothesis can now be established which in this case will be linkage. In reexamining the data in Table 3.3, two classes occur with a very high frequency and two with a very low frequency. As indicated earlier, the high frequency classes are the result of parental (noncrossover) gametes, while the low frequency classes result from crossover events. It is possible to calculate from the data the approximate distance in terms of crossover probability between these two loci. Note that distance is calculted as probability of a crossover occurring, rather than the actual physical distance that may lie between the two genes, although these two distance concepts may be in close agreement. The distance is computed from the testcross data in the following manner.

$$\text{Crossovers units} = \left(\frac{\text{Crossover progeny}}{\text{Total progeny}}\right) \times 100$$

This formula simply compares the number of gametes that have been produced as a result of a crossover with the number of the gametes in the population. As the distance between any two loci increases, the probability of a crossover occurring between these

Table 3.3 Sample Testcross Data With Linkage

Phenotype	O	C	$(O-C)^2$	$\frac{(O-C)^2}{C}$
Resistant yellow	43	30	169	5.63
Susceptible green	49	30	361	12.03
Resistant green	16	30	196	6.53
Susceptible yellow	12	30	324	10.80
	120			34.99

two loci also increases and thus the number of crossover gametes increases. Loci that are very close together have little likelihood of crossovers between them and crossover gametes are produced in very low frequencies. On the other hand, genes with large distances between them will produce a high proportion of crossover gametes. In our example the distance is calculated as $[(16+12=28)/120]\ 100 = 23.3$ crossover units.

The maximum distance that can be measured, using only two loci, is less than 50 crossover units. This is because a crossover occurring between two loci in every meiotic cell will produce the four types of gametes with equal frequency and the testcross data will show a distance of 50 crossover units. The data are indistinguishable from those of independent assortment where all four gametes are also produced with equal frequency. Linkage maps longer than 50 units are constructed through the use of additional loci. For example, if A is more than 50 units from C, but A and B are linked with less than 50 units, and B and C are also linked with less than 50 units, it follows that A and C must also be on the same chromosome. The distance between A-C is approximated by the two values of A-B and B-C. This is called a three-point linkage map and not only measures the distance between the three loci but can also determine the sequence of the loci on the chromosome. By combining the results of many experiments, linkage maps can be constructed. Detailed maps are available in corn, tomato, and barley, for example. The map of barley showing the known location of loci on each chromosome is given in Figure 3.6.

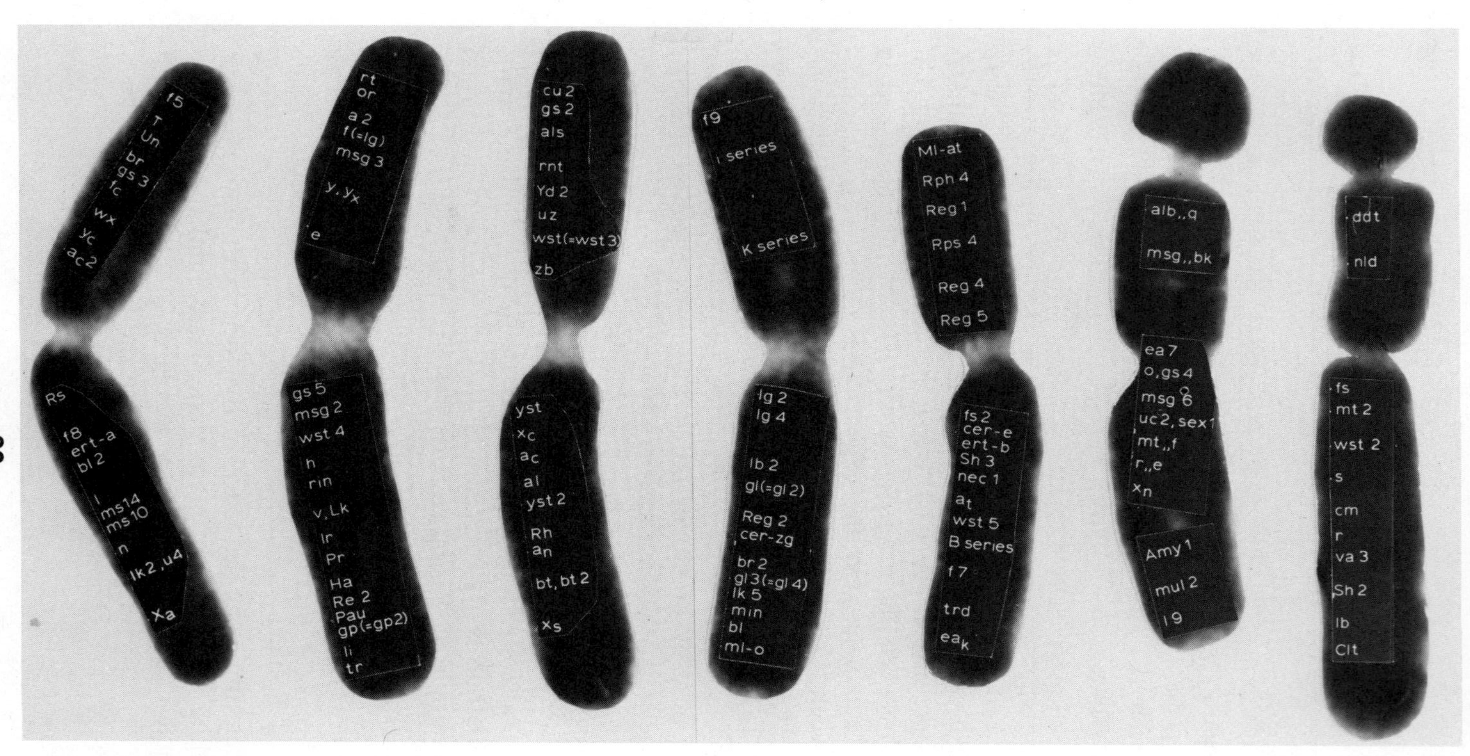

Figure 3.6. Loci positions on the seven chromosomes of barley. The chromosomes are identified from I to VII starting from the left. Symbols for the different genes measured on each chromosome have been superimposed on each arm. The centromeres lie directly across the center of the plate. Chromosomes VI and VII each have a secondary constriction on the upper arm. (Courtesy T. Tsuchiya, Colorado State Univ.)

Linkage can also be detected by studying F_2 ratios for deviations from normal independent assortment. The data presented in Table 3.3 show that the four gametes from the F_1 are produced with average frequencies of 0.383 *RY*, 0.383 *ry*, 0.116 *Ry*, and 0.116 *rY*. These values can be used to produce the Punnett square presented in Figure 3.7. The observed F_2 genotypic and phenotypic frequencies deviate from those expected with independent assortment. Based on an independent assortment ratio of 9 : 3 : 3 : 1, the phenotype frequencies in percent should be 56 : 19 : 19 : 6. The calculated chi-square value for the data in Figure 3.7 would be 23.17. The tabular value with three degrees of freedom indicates that this ratio is a very poor fit to a 9 : 3 : 3 : 1. Here, as in testcross data, more parental type gametes occur than are expected. Deviations in the F_2 data are more difficult to analyze than those in testcross data, but breeders often have only F_2 populations, especially where homozygous recessive tester parents are not available. Deviations from expected F_2 ratios based on independent assortment may result from linkage, but they may also occur from many other genetic causes. Linkage is most often determined

	.383 *RY*	.116 *Ry*	.116 *rY*	.383 *ry*
.383 *RY*	.147 *RRYY*	.044 *RRYy*	.044 *RrYY*	.147 *RrYy*
.116 *Ry*	.044 *RRYy*	.013 *RRyy*	.013 *RrYy*	.044 *Rryy*
.116 *rY*	.044 *RrYY*	.013 *RrYy*	.013 *rrYY*	.044 *rrYy*
.383 *ry*	.147 *RrYy*	.044 *Rryy*	.044 *rrYy*	.147 *rryy*

Ratio (in percent)	
Resistant yellow	: 64.3
Resistant green	: 10.1
Susceptible yellow	: 10.1
Susceptible green	: 14.7

Figure 3.7. An F_2 Punnett square produced with gamete frequencies derived from Table 3.3. The linkage condition has dictated that each gamete does not occur with a 0.25 frequency.

by the abnormally high frequency of paired character expressions in segregating populations. The exact distances may be unknown. In plant breeding, linkage plays an important role since blocks or groups of genes tend to move together. The advantages and problems of linkage are covered in breeding chapters.

We have considered the situation where the alleles at both loci on one chromosome are in the dominant condition and the alleles at both loci on the other chromosome are recessive. This is called coupling and results in parental gametes containing both dominant and both recessive alleles with crossover gametes containing a dominant allele at one locus and a recessive at the other. However, the reverse situation, called repulsion, can be encountered where the dominant allele at one locus is linked with the recessive allele at the other. In this case the parental gametes each contain a dominant at one locus and a recessive at the other (Fig. 3.4*d*), while the crossover gametes have either both dominant or both recessive alleles (Fig. 3.4*e*).

The question is often raised regarding the relationship of linkage to the data generated by Mendel. He studied seven characters that appeared to be independently assorted from one another. Peas have seven chromosome pairs, which leads to the obvious speculation that Mendel was lucky enough to have selected seven characters each located on a different chromosome pair. In fact, subsequent chromosome analysis of the pea plant has shown that he chose two loci on each of two chromosomes and one on each of three others. This does not invalidate his data analysis, however, because of the distance factor that allows two loci separated by a large distance to appear as if they were independently assorted. Instead of being extremely lucky, Mendel chose some loci that were widely separated in linkage distance on the same chromosomes.

SUMMARY AND COMMENTS

The beautifully orchestrated chromosome mechanisms provide the mechanical means of transmitting genetic information from generation to generation on a highly repeatable basis. The fact that loci are linearly arranged in the chromosome is very important in plant breeding programs since blocks of genes or alleles are inherited together with a much higher frequency than if they were independently assorted. This may be good if only desirable alleles are physically associated but it can be very undesirable if detrimental and valuable alleles are linked.

I have found linkage very valuable in at least one instance where disease resistance was associated with red color for plant type. The evaluation of resistance was difficult because of the inability to consistently produce the disease on segregating populations. The red color showed up very regularly, however, and so it was possible to maximize resistant plants in our populations by selecting those plants that had red color phenotype. Unfortunately, the relationship was not perfect because recombinations through crossover did occur periodically so that some red plants were susceptible, while some white plants were resistant. We did use the linkage condition to bias our populations in our favor however, which is often as much as a breeder can ask when dealing with difficult traits.

QUESTIONS

1. Two true breeding parents, one tall and the other short, were crossed to produce several F_1 plants. These were allowed to intermate and the F_2 generation gave a 691 tall : 217 short ratio. What is the chi-square probability of a single locus segregation? **Answer:** 0.1–0.5.

2. Suppose the same conditions exist as in Question 1, but in another experiment the F_2 segregation is 654 tall : 254 short. What is the chi-square probability of a single locus segregation? **Answer:** 0.01–0.05.

3. Of the tall F_2 segregates in Question 2, 203 produce all tall progeny and 451 give segregating progeny for tall and short. Calculate the chi-square probability of these results supporting a single locus segregation hypothesis. **Answer:** 0.10–0.50.

4. An individual with a chromosome number of $2\underline{n} = 6$ is heterozygous for three independently assorting loci. Use gene symbols *A, a, B, b,* and *C, c.* What are the possible gametes that this individual can produce? **Answer:** *ABC, ABc, Abc, AbC, aBC, ABc, abC, abc.*
Diagram all anaphase I and anaphase II events placing the alleles on the chromosomes to produce the above gametes.

5. Disease resistance is dominant (R) over susceptibility (r). Yellow (Y) color is dominant over green (y). The testcross between a homozygous recessive and a heterozygous F_1 produces the following ratio:
Resistant yellow 153
Susceptible yellow 426
Resistant green 417
Susceptible green 164
   a. What is the calculated chi-square value based on an independent assortment assumption? **Answer:** 238.9
   b. Is independent assortment likely? **Answer:** No.
   c. Assuming linkage, which of the testcross classes are the result of crossovers? **Answer:** Resistant yellow and susceptible green.
   d. What is the map distance between these two loci? **Answer:** 27.3 units.
   e. What were the genotypes and phenotypes of the two true breeding parents crossed to produce the F_1? **Answer:** *RRyy* (resistant green) X *rrYY* (susceptible yellow).
   f. Is coupling or repulsion present? **Answer:** Repulsion.

4

GENES

In Chapter 3 it was pointed out that genes are actually segments of DNA arranged in a linear sequence within each chromosome. Each gene site or locus is potentially capable of transmitting information from the DNA through RNA to enzymes responsible for the biological mechanisms controlling growth, function, and characteristics of the organism. The process of differentiation within each individual is accompanied by the switching on or off of specific genes in each organ. While each somatic cell normally has a complete complement of genetic information, only that part pertinent to organ development and function is used at any given time.

A great deal of scientific effort has been expended in recent years to more completely describe and understand the structure and function of genes, especially in lower organisms. For our purposes it is sufficient to recognize the genes as subdivisions of the chromosomes that provide information to the functioning mechanisms responsible for character expression.

GENE VARIABILITY

In Chapter 2 the concept of alleles, different forms of the same gene, was mentioned. Mendel, in his studies, used two allelic variations for each character. For example, height was governed by a locus for which two alleles were available. He designated them as dominant and recessive and identified them with capital and lower case letters. The dominant allele produced tall individuals, while the recessive allele, when homozygous, resulted in short or "dwarf" plants. His terminology implied that only two allelic forms of each gene exist. This may be true for some loci, but we now know that for many loci several alleles may be present in a population of individuals. The presence of several allelic forms for one locus is called a multiple allele series. The existence of multiple alleles requires a new designation system. The most common method uses superscript or subscript numbers or letters associated with the gene symbol. For example, if a multiple allele series exists for the gene *A*, then the series might be designated $A_1, A_2, A_3 \ldots$ or $A^1, A^2, A^3 \ldots$. Notice that the condition of dominance or recessiveness is not always indicated by the new symbolism. Although dominance and recessiveness may exist in a multiple allele series, we will see later in this chapter that other types of gene action can also occur. Any individual with one chromosome pair containing the *A* locus will have no more than two alleles of this gene that are identical in a homozygote but different in a heterozygote. Although any one individual may be limited to a maximum of two different allelic forms, a multiple series can exist in a population because different individuals can carry any two of the possible alleles.

Many multiple allele series occur in crop plants. An interesting example is described in sorghum by Quinby (11) where the character of days from planting to flowering is under the genetic control of a multiple allele series at each of four loci. At the first locus 2 dominant and 11 recessive alleles have been described, at the second there are 12 dominant and 2 recessive alleles, at the third 9 dominant and 7 recessive, and 11 dominant and 1 recessive at the fourth locus. All four loci are designated *Ma* to indicate maturity. Subscript numbers identify each of the four loci. Superscript abbreviations indicate the allelic variation by source. For example, $Ma^{F}{}_{1}$ indicates that this is a dominant allele variation at locus one and the allele has been obtained from a particular genetic source identified by F. This, then, is a case of multiple alleles at each of several loci. In the inheritance system of this character, several varieties may have the same general genetic combination of dominant and recessive alleles at the four loci but different allelic forms of dominants and recessives can result in flowering date differences.

Allelic variations arise from mutations that alter the information within the DNA sequence of the gene. A mutation could be considered an error in the inheritance information mechanism. The word "error" implies that all mutations—and consequently new alleles—are undesirable. While this is true in many cases, it is important to remember that some alleles produced in the mutational process, either alone or in combination with other loci, may have some advantages for the plant. In fact, mutations are the source of genetic variability that provides the plant breeder with the raw materials for new and potentially better genetic combinations.

GENE ACTION

Our common concept of the way genes work or express themselves (gene action) has been one of dominance or recessiveness. The allele either expresses itself completely in the phenotype, or is not expressed at all. This concept was adequate for Mendel's studies but would not satisfy later investigations where additional phenotypes appeared in the F_2 or where the F_1 did not have the exact phenotype of the dominant parent. Research has shown that there are many different kinds of gene actions and interactions that account for segregation patterns differing from those obtained by Mendel. As we describe a few of these variations, remember that the basic assumptions of pairs and randomness are not being violated. Instead, additional refinements are being imposed on the existing mechanisms.

The types of gene action can be broken down into the two general categories of interaction: between alleles at the same locus, (intralocus), and interaction between alleles at different loci (interlocus). While some situations fit into one of these categories perfectly others tend to overlap the two divisions.

Intralocus Interactions

There are three general types of intralocus interaction. Dominance is typified by Mendel's study where the normal phenotypic segregation ratio in the F_2 from two homozygous parents is 3

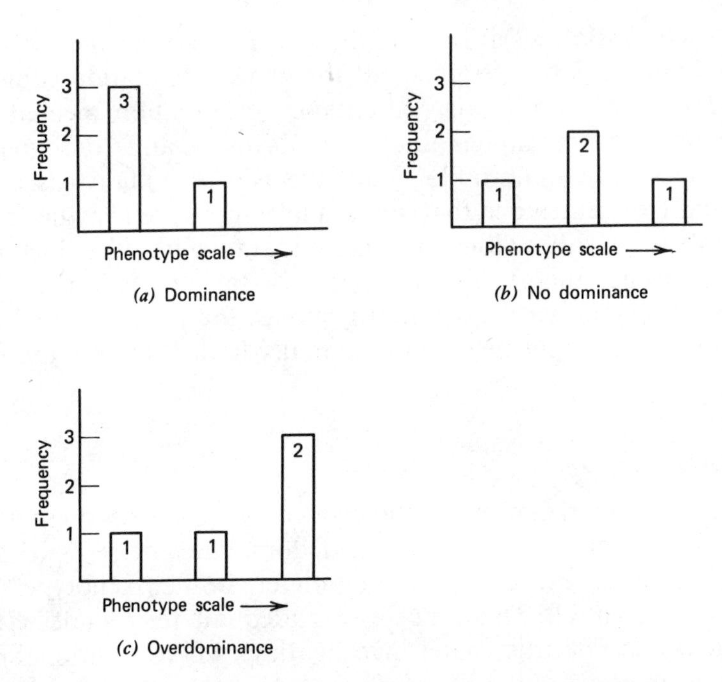

Figure 4.1. F_2 frequency distributions for different intralocus interactions.

dominant : 1 recessive. A typical F_2 frequency distribution is shown in Figure 4.1*a*.

With a second type, called no dominance, or additive, the phenotype of the heterozygote falls midway between that of the two homozygous parents (Figure 4.1*b*). In the four-o'clock plant a cross between red and white flowered parents will produce a pink flowered F_1. The F_2 segregation will be 1 red : 2 pink : 1 white. In the completely additive system the heterozygote will have a phenotypic value exactly midway between the two homozygotes. Variations of this system can occur in which the heterozygote may achieve a value closer to one parent than the other. This is termed partial dominance. In a cross of the cultivated strawberry with the wild species (10,12) the F_1's were nearly as early in flowering as the early parents and the F_2 segregation patterns confirmed partial dominance for earliness.

Overdominance is the third general category of intralocus interaction. In this situation the heterozygote has a phenotypic value outside the range between the two parents as indicated in Figure 4.1*c*. This type of gene action is of particular interest to the hybrid plant and animal industries because of productive potentials associated with hybrid vigor. Because of the heterozygosity, overdominance cannot be permanently fixed in a homozygous condition. The issue of overdominance is discussed further in Chapter 15 on hybrids and heterosis.

Interlocus Interaction

Just as different types of allelic interaction can occur within a locus, many types of interaction can take place between loci. We will consider several examples of interlocus interaction on character expression resulting in F_2 distribution pattern changes. The expression of one allele may be changed by the presence or absence of an allele or alleles at another locus. This is called epistasis and requires that at least two loci operate in the expression of a single character. If the independent assortment assumption is satisfied, variations in normal dihybrid ratios can be expected. Several examples are provided to illustrate epistatic effects on segregation patterns.

In soybeans the pod wall loses its green color at maturity and takes on a color characteristic of the variety, ranging through various shades of black, brown, and tan. Bernard (3) has shown that two loci control the three colors. At one locus, dominant L_1 causes black pods. Recessive l_1 produces either brown when dominant L_2 is present at the second locus or tan when recessive l_2 is homozygous. A cross of $L_1L_1L_2L_2$ (black) $\times$ $l_1l_1l_2l_2$ (tan) would produce an F_1 of $L_1l_1L_2l_2$, which would be black. The F_2 segregation pattern would be:

9 L_1 __ L_2 __	black
3 L_1 __ l_2l_2	black
3 $l_1l_1L_2$ __	brown
1 $l_1l_1l_2l_2$	tan

The blank at each locus can be filled in with either a dominant or recessive allele, since dominance occurs within each locus. In this example, the L_1 allele is overriding (epistatic) to the L_2 allele that

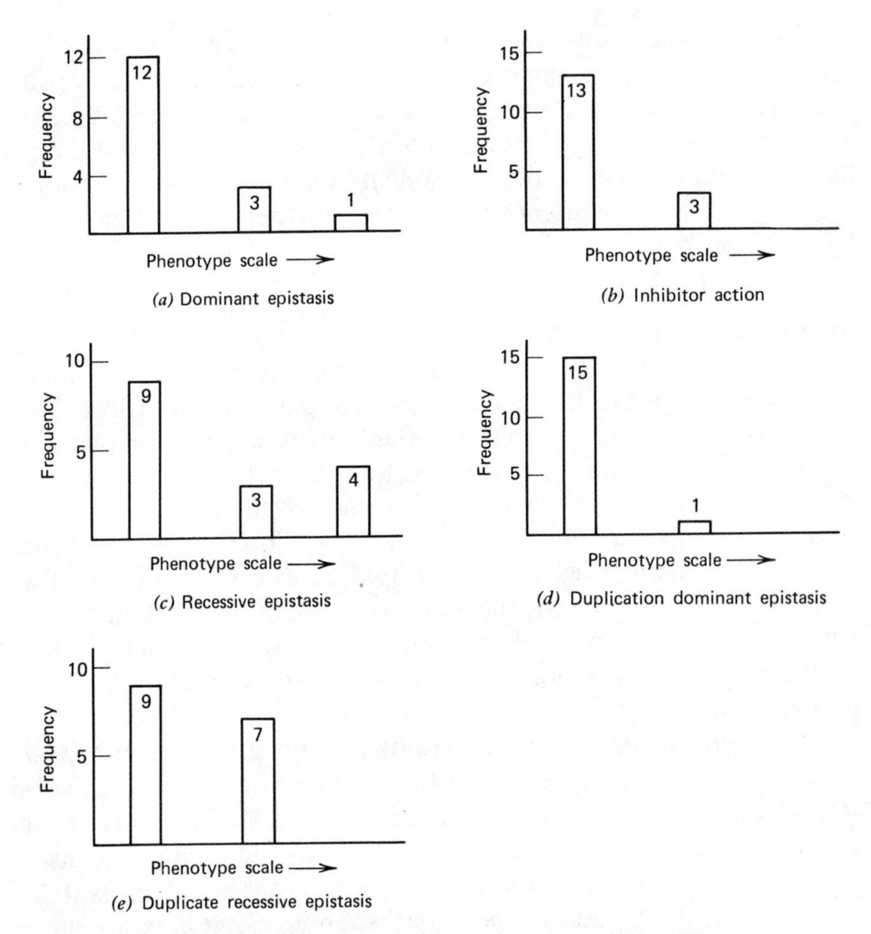

Figure 4.2. F_2 frequency distributions for different interlocus interactions.

is being overridden (hypostatic). This is called dominant epistasis, sometimes referred to as masking gene action, and results in a 12 : 3 : 1 F_2 ratio as shown in Figure 4.2*a*.

A variation of dominant epistasis involves the inhibitor type of mechanism. In corn a dominant inhibitor allele, *I*, is epistatic to the dominant endosperm color producing allele, *C*, at another locus. The presence of the dominant *I* allele will inhibit any color formation regardless of the allelic condition at the color locus. The F_2 segregation pattern is as follows:

9 $C\_\ I\_$	colorless
3 $C\_\ ii$	colored
3 $cc\ I\_$	colorless
1 $cc\ ii$	colorless

This system will produce a 13 color : 3 colorless distribution illustrated in Figure 4.2*b*. These allelic variations are described by Neuffer et al. (8).

Flower color in sunflowers is controlled by two independently assorted loci with multiple alleles and dominance within each locus. Fick (5) demonstrated that in crosses of yellow (*LL LaLa*) × lemon (*ll lala*) the F_1 was yellow and the F_2 segregated in a ratio of 9 yellow : 3 orange : 4 lemon. The following is the basis of the F_2 distribution:

9 $L\_\ La\_$	yellow
3 $ll\ La\_$	orange
3 $L\_\ lala$	lemon
1 $ll\ lala$	lemon

This is a case of recessive epistasis, often called modifying gene action, where the homozygous recessive at one locus is epistatic to the dominant at another locus. The frequency distribution is illustrated in Figure 4.2*c*.

A chlorophyll controlling genetic system, described by Bernard and Weiss (4), exists in soybeans where at one locus the dominant allele *G* produces normal green seed coat color and the recessive produces yellow. At a second locus the dominant Y_3 allele results in green color and the recessive y_3 causes the leaves to turn yellow with age. In a cross between two green parents, $GG\ y_3y_3 \times gg\ Y_3Y_3$, the F_1, $Gg\ Y_3y_3$, will be green. The F_2 will be distributed in a 15 green : 1 yellow pattern based on the following:

9 $G\_\ Y_3\_$	green
3 $gg\ Y_3\_$	green
3 $G\_\ y_3y_3$	green
1 $gg\ y_3y_3$	yellow

The only chlorophyll deficient plants are those that are homozygous recessive at both loci. This type of allelic action, called duplicate dominant epistasis, is the condition where the dominant allele at either locus can override the homozygous recessive at the other locus. The frequency distribution for duplicate dominant epistasis is shown in Figure 4.2*d*.

In Ladino clover, Atwood and Sullivan (2) described the inheritance of hydrocyanic acid (HCN) content. In crosses of high × low parents the F_1 plants were high in HCN and the F_2 segregation produced a 9 high : 7 low ratio. This is a case of duplicate recessive epistasis, or complimentary gene action, in which the homozygous recessive condition at either locus prohibited the dominant expression at the other locus. The F_2 distribution composition is given as follows:

9 $A\_\ B\_$	high
3 $aa\ B\_$	low
3 $A\_\ bb$	low
1 $aa\ bb$	low

The F_2 frequency distribution for duplicate recessive epistasis is given in Figure 4.2*e*. This example also serves to illustrate the physiological implications of genes. Atwood and Sullivan explain the genetic results on the basis that the dominant allele at one locus produces the substrate cyanogenic glucoside, while the dominant allele at the other locus produces the enzyme that releases HCN from the glucoside. If either the substrate or the enzyme is missing, the HCN will not be produced. The recessive allele at each locus represents an alternative form of DNA information with a different biochemical and subsequent phenotype result. The operation and expression of all genetic systems can be explained in some physiological manner but at the present time we are generally limited to interpretation based on phenotypic observations. As research continues we will come to know more about the basic functions of genetic control mechanisms, which in turn will have special implications on physiological manipulations through breeding.

ADDITIVE GENE ACTION

Thus far we have considered interlocus interaction in terms of epistasis where an allele at one locus may be affected by the presence of an allele at another locus. Another very common type of gene action, which is not epistatic by our definition, is the case of additivity where each allele at one locus will add or subtract an increment of phenotypic value. One classic example of this type of gene action is provided by the inheritance of kernel color in wheat first described by Nilsson-Ehle (9). He found that color is governed by three loci R_1, R_2, and R_3 independently assorted with two alleles at each locus. Kernel color ranged from very dark red to white, and the intensity of color depended on the number of color-adding alleles present in the genotype. While there was some dominance within each locus, the gene action was primarily additive both within and among loci. The very dark red parent was $R_1R_1R_2R_2R_3R_3$, while the completely white parent was $r_1r_1r_2r_2r_3r_3$. The F_1, $R_1r_1R_2r_2R_3r_3$, was intermediate in color. A series of color classes appeared in the F_2 with a nearly continuous normal distribution between the two extreme phenotypes. The F_2 frequency distribution pattern is given in Figure 4.3. The F_2 performance is explained by the fact that all genotypes containing the same number of color-adding alleles would have the same phenotype. The phenotypic composition of the F_2 population can be obtained by the expansion of the formula $(a + b)^n$ where a = the alleles adding value, b = the alleles with no additive value, and n = the total number of alleles, which in our case is six. The expansion of $(a + b)^6$ produces a formula of $a^6 + 6a^5b + 15a^4b^2 + 20a^3b^3 + 15a^2b^4 + 6ab^5 + b^6$. We find that there are seven groups of genotypes and corresponding phenotypes with the frequency of 1 : 6 : 15 : 20 : 15 : 6 : 1. The duplication of phenotypic results by different genotypes with additive gene action can be illustrated by considering the term $20a^3b^3$, which indicates that there are 20 individuals with 3 additive alleles (a^3) producing intermediate color. This term includes the following genotypes and their frequencies:

$R_1R_1R_2r_2r_3r_3$	2
$R_1R_1r_2r_2R_3r_3$	2
$R_1r_1R_2R_2r_3r_3$	2
$r_1r_1R_2R_2R_3r_3$	2
$R_1r_1r_2r_2R_3R_3$	2
$r_1r_1R_2r_2R_3R_3$	2
$R_1r_1R_2r_2R_3r_3$	8
	20

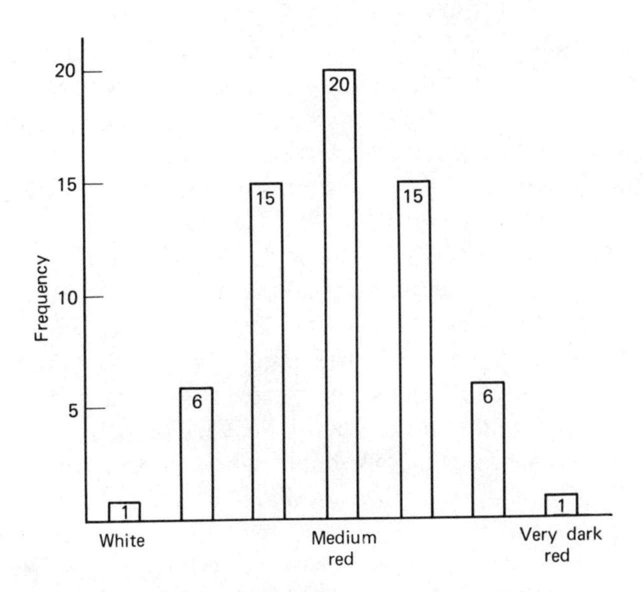

Figure 4.3. F_2 frequency distribution for additive kernel color inheritance in wheat with three loci independently assorted.

Note that each of these genotypes has three alleles that will add an increment of color to the kernel. Inheritance such as that illustrated by the wheat seed color system is often called multiple factor inheritance, meaning that several loci are involved in the genetic control with an associated loss of simplicity in inheritance patterns. Further variations caused by additivity are discussed in more detail in Chapter 8.

TRANSGRESSIVE SEGREGATION

Using the same genetic example, consider the cross of $R_1R_1R_2R_2r_3r_3 \times r_1r_1r_2r_2R_3R_3$. One parent will be dark red and the other will be very light red. The F_1 will be intermediate in color, having three color alleles, and the F_2 will give a complete range of segregation from very dark red to completely white. In this cross some of the segregating progeny fall outside the phenotypic limits of the dark red and very light red parents. This is called transgressive segregation and, in our example, results from the additive nature of the gene action. The same effect can be achieved with other types of allelic interactions including epistasis and overdominance. For example, the case cited for duplicate dominant epistasis in soybeans also illustrated transgressive segregation because some of the progeny from two green parents had a yellow phenotype. Transgressive segregation is of considerable importance to the breeder because it allows for the possibility of obtaining segregates that are better (or worse) than either parent. The extent to which this condition can be exploited depends primarily on the type of gene action involved—as we will see in subsequent discussions on various breeding schemes in crops with differing population dynamics.

MINOR MODIFIERS

A final type of interlocus interaction involves what are called minor modifying alleles. These are important because they may result in slight deviations from expected phenotypic values associated with genes having major impact on the phenotype. Minor modifiers are usually present to a varying extent in most inheritance systems. They are particularly evident in disease resistance mechanisms where major genes condition well-defined reaction phenotypes. Slight fluctuations within the general phenotypic classes often occur because of minor modifying genes. The value of minor modifiers in breeding programs depends on the number of modifiers present, the extent of their influence on the character, and the direction in which they move the phenotypic value.

PENETRANCE AND EXPRESSIVITY

A compounding problem in gene action sometimes occurs when identical genotypes will not express the same phenotype even with environmental uniformity. We call this a difference in penetrance or the frequency with which a gene shows a particular effect. For example, in tomatoes Alon et al. (1) described the genetics of resistance to the disease *Fusarium* wilt. Two varieties homozygous for a dominant resistant allele, *I,* produced low numbers of susceptible seedlings when exposed to the disease. Even though both varieties had the same genotype, *II,* they gave different numbers of susceptible plants under identical environments. This means resistance did not have 100 percent penetrance and penetrance differed between varieties. In addition, the alteration of the environment in terms of inoculum concentration or soil temperature changed the penetrance level in each population. The reasons for less than 100 percent penetrance are not known but the interaction of genes and environment undoubtedly plays a major role.

Another situation that tends to obscure clear-cut genetic results involves expressivity or the degree of gene expression. In the case of the tomato wilt resistance system, variation within the susceptible reaction of susceptible plants would reflect a difference in the expressivity of the gene.

While many genes have no variation in their expression under a single environment, others may vary considerably. It is possible that minor modifying genes differing from individual to individual affect the major genes, in which instance selection for greater or lesser penetrance or expressivity should be possible. This occurred with the inheritance of fasciation, the fusion of branches in peas. Marx and Hagedorn (7) identified variation for both penetrance and expressivity in this character but were able to select lines with complete penetrance and uniform expressivity. This was probably the result of isolation and stabilization through selection of a single genotype for all loci affecting the trait.

PLEIOTROPY

In some cases a gene may have what appears to be an effect on two or more separate and distinct characters. This is called pleiotropy or a pleiotropic effect. Habans and Dahiya (6) reported a recessive allele produced by mutation in chickpeas that controlled leaf length, leaflet size, plant height, number of branches, pod number, and seed number. The pleiotropy, while extreme in this case, may be the result of the interruption or altering of a single physiological mechanism involved in all these characters. In other cases, what was initially thought to be pleiotropy was actually the result of very closely linked loci. Studies with large numbers of progeny produced a low frequency of recombinant genotypes, voiding the pleiotropy hypothesis.

SUMMARY AND COMMENTS

A wide array of interactions can occur within and between loci. Although these do not invalidate Mendelian principles, they do add complexity to inheritance patterns. They also offer additional opportunities for unpredicted results through plant breeding manipulations.

Some of my most productive and satisfying breeding achievements have come about because of transgressive segregation. I have often made crosses between what appeared to be mediocre parents with the startling emergence of rather excellent plant types among the progeny. Until we completely understand the mechanics and control of inheritance, and can predict genetic results with unfailing accuracy, the opportunity will exist for unexpected exciting genetic creations.

REFERENCES

1. Alon, H., J. Katan, and N. Kedar. 1974. Factors affecting penetrance of resistance to *Fusarium oxysporum* f. sp. *lycopersici* in tomatoes. *Phytopathology* 64:455-461.

2. Atwood, S. S., and J. T. Sullivan. 1943. Inheritance of a cyanogenetic glucoside and its hydrolyzing enzyme in *Trifolium repens*. *J. Hered.* 34:311-320.

3. Bernard, R. L. 1967. The inheritance of pod color in soybeans. *J. Hered.* 58:165-168.

4. ———, and M. G. Weiss. 1973. Qualitative genetics. In B. E. Caldwell (ed.), Soybeans: improvement, production, and uses. *Am. Soc. Agron. Monograph* 16:117-154.

5. Fick, G. N. 1976. Genetics of floral color and morphology in sunflowers. *J. Hered.* 67:227-230.

6. Habans, S., and B. S. Dahiya. 1974. A note on inheritance of busy mutant in chickpea. *Current Sci.* 43:731-732.

7. Marx, G. A., and D. J. Hagedorn. 1962. Fasciation in pisum. *J. Hered.* 53:31-43.

8. Neuffer, M. G., Loring Jones, and Marcus S. Zuber. 1968. *The Mutants of maize.* Crop Sci. Soc. of Am., Madison, Wisc.

9. Nilsson-Ehle, H. 1911. Kreuzungsuntersuchungen an Hafer und Weizen. *Lunds Univ. Arsskr.* N. F. Afd. 2 Bd.5. Nr.2. 1-122.

10. Powers, L. 1945. Strawberry breeding studies involving crosses between the cultivated varieties (*Fragaria* × *ananassa*) and the native Rocky Mountain strawberry (F. ovalis). *J. Agric. Res.* 70:95-122.

11. Quinby, J. R. 1967. The maturity genes of sorghum. *Adv. Agron.* 19:267-305.

12. Scott, D. H., A. D. Draper, and L. W. Greeley. 1972. Interspecific hybridization in octaploid strawberries. *Hort. Science* 7:382-384.

QUESTIONS

1. Assume two loci, independently assorted, governing a character. $A_1A_1B_1B_1$, has a value of 4. A_2 is completely dominant over A_1 and adds 8 units. B_2 is completely additive to B_1 and adds 2 units. Additivity occurs between loci. A cross is made between $A_1A_1B_1B_1$ and $A_2A_2B_2B_2$.
   a. What is the value of the second parent? **Answer:** 16.
   b. What is the value of the F_1? **Answer:** 14.
   c. What is the F_2 phenotypic segregation ratio? **Answer:**

Freq.	*Value*
3	16
6	14
3	12
1	8
2	6
1	4

2. Using the information in Question 1, what would be the ratios from the following crosses?
   a. $A_1A_2B_1B_1 \times A_1A_2B_2B_2$ **Answer:**

Freq.	*Value*
3	14
1	6

   Is this transgressive segregation: **Answer:** Yes.
   b. $A_1A_2B_2B_2 \times A_2A_2B_2B_2$
   Answer: All with a value of 16.
   Is this transgressive segregation: **Answer:** No.

5

CHROMOSOME NUMBERS

The reliability of inheritance systems is based in large part on chromosome number stability and repeatability. Mitosis and meiosis are beautifully orchestrated chromosome duplication, division, and movement mechanisms that contain nearly foolproof safeguards against the changes in chromosome numbers. Limited opportunity does exist, however, for the occurrence of variations in numbers of chromosomes. Although this is generally a low frequency event when compared to the highly repeatable normal cell division happenings, it does offer another genetic raw material potential for evolution and improvement.

Cytological studies have produced a wealth of information on chromosome numbers in many plant and animal species. Publications compiling chromosome numbers in the plant kingdom include the *Chromosome Atlas of Flowering Plants* by Darlington and Wylie (4), and *Chromosome Numbers of Central and Northwest European Plant Species* by Löve and Löve (6). Both books represent the accumulated results of numerous cytological investigations, and considerable additional literature on chromosome numbers becomes available each year.

We will consider two chromosome number concepts. The first, euploidy, involves sets or groups of chromosomes called genomes. The second, aneuploidy, deals with variations in one or a small number of chromosomes. There are many complicated variations within and between both types of ploidy. Our discussion is limited to the ploidy terminology and ideas related to plant breeding techniques.

EUPLOIDY

Euploidy is defined as the variation in the numbers of basic sets of chromosomes called genomes. A "basic set" implies the smallest number of chromosomes ever having existed in the evolution of a species. For example, it appears that the smallest number of chromosomes contained within a set or genome in cotton is 13. Seven chromosomes within each genome is common for many of the grasses including wheat, barley, and oats. By contrast, the alfalfa genome contains eight chromosomes, as does the apricot. Table 5.1 lists the number of chromosomes within the genome of several commercially important crop species and also indicates that some of the species possess several genomes.

Table 5.1. Chromosome Number and Ploidy Conditions in Several Commercially Important Crop Species

Common Name	Chromosome Number in X	2n	Ploidy
Alfalfa	8	32	4X
Almond	8	16	2X
Apple	17	34	2X
Apricot	8	16	2X
Barley, cultivated	7	14	2X
Cherry, sour	8	32	4X
Cherry, sweet	8	16	2X
Cotton, asiatic	13	26	2X
Cotton, upland	13	52	4X
Oats, cultivated	7	42	6X
Orange, sweet	9	18	2X
Peach	8	16	2X
Pear	17	34	2X
Potato	12	48	4X
Strawberry, cultivated	7	56	8X
Tobacco, cultivated	12	48	4X
Wheat, bread	7	42	6X
Wheat, durum	7	28	4X

"X" is employed as the general genome label to identify the number of genomes present. Thus, 1X, 2X, or 3X individuals would possess one, two, or three genomes. To find the total number of chromosomes the number of X's or genomes must be multiplied by the number of chromosomes contained within the genome.

The general terminology of euploidy is quite simple. An individual with a single set or genome (1X) is called a monoploid. A diploid has two genomes (2X). A triploid is 3X, a tetraploid—4X, a pentaploid—5X, a hexaploid—6X, and so on. The euploidy levels above diploids are generally grouped into a class called polyploids. The ploidy levels of several commercial plant species are given in Table 5.1. For example, cultivated strawberries are octaploid, alfalfa is tetraploid, and the pear is diploid.

A distinction must be made between the identification of the genome number (X) and the chromosome number ($\underline{n}$). The X designation is always reserved for the genome, which in turn contains a specific number of chromosomes. The somatic chromosome number of any individual is identified as $2\underline{n}$ and the gamete or haploid chromosome number is $\underline{n}$. In some cases X and $\underline{n}$ are equal; in others they are not. A diploid individual with a genome constitution of 2X, where X contains 7 chromosomes has a total chromosome number of 14, or 7 pairs, which results in a $2\underline{n} = 14$ or $\underline{n} = 7$ designation. In this case $2X = 2\underline{n} = 14$. However, a hexaploid individual would have a 6X designation but would still have a $2\underline{n}$ somatic chromosome number and $\underline{n}$ gamete chromosome number. Thus, in the hexaploid, $6X = 2\underline{n} = 42$ and $\underline{n} = 21$. Special cases arise when uneven numbers of genomes occur. For example, in a triploid (3X) individual where X contains 7 chromosomes, the $2\underline{n}$ or somatic chromosome number is equal to 21, but the $\underline{n}$ number varies because of unequal chromosome segregation during meiosis.

Autopolyploids (auto meaning self) contain genomes that are identical in chromosome makeup. For example, an autotriploid contains three genomes (3X) and all genomes contain exactly the same set of chromosomes. Autopolyploidy can be produced in several different ways. Occasionally in nature, gametes are generated that have not undergone normal chromosome reduction in meiosis. They are called unreduced gametes and are most commonly produced by meiotic abnormalities where reduction division is initiated but is arrested in late anaphase or some stage prior to the formation of two new cells. The result is a gamete that contains the full number of chromosome pairs ($2\underline{n}$). Should this gamete combine with one produced by normal meiosis, the resulting zygote will have an extra or triple set of chromosomes. Higher levels of autopolyploidy can be achieved by repeating the duplication of genome numbers, but difficulties in plant viability and vigor are normally experienced as chromosome numbers go beyond reasonable levels. The critical level varies with the plant species and while some species can tolerate autooctaploidy, autotetraploidy is unacceptable to others.

In addition to natural means, polyploidy can be induced artificially with the use of chemical compounds or by special heat or cold treatments. Of the chemical compounds, by far the most popular and commonly used is colchicine, which is applied to active meristematic tissue to interrupt the spindle fiber formation and completion of cell division. The resulting cells are doubled in chromosome number and if the chemical is applied for prolonged periods, repeated multiplication of chromosome numbers will occur. More discussion on artificially induced polyploidy is provided in Chapter 18 on chromosome manipulations.

One of the serious problems associated with autopolyploidy involves the strong tendency for homologous chromosomes to pair during meiosis. In a diploid condition, synapsis results in even chromosome distribution to the gametes. In an autopolyploid, however, the chromosomes tend to associate in groups of more than two. For example, in an autotriploid three of each chromosome synapse in various manners along their length in meiosis I. Although no more than two chromosomes will pair at any one point, all three will be involved in the total synaptic complex. As they begin to move, the distribution of chromosomes to the poles occurs in a more or less random manner. This results in some gametes receiving one of the chromosomes, some receiving two, and very infrequently some receiving none or all three. Since this same thing is happening simultaneously with all the chromosomes in the genome, production of gametes with unbalanced chromosome numbers is a highly common occurrence. Normally, these gametes do not function properly in the pollination and fertilization process and thereby cause a high degree of sterility.

Allopolyploids (allo meaning different) contain genomes from different genera, species, or subspecies. Specific genomes must be identified so we can follow their role in the formation of polyploids. Genome identification is commonly done by the use of uppercase letters. The three genomes in wheat, for example, have been labeled as A, B, and D. Each genome contains seven chromosomes, but those of the A genome are not homologous with those of B or D. Wheat represents a classic example of the evolution of an allopolyploid series and is summarized by Sears (9). The wheat used in today's commercial bread industry is an allohexaploid (6X) with a genome formula of AABBDD. This translates into $2\underline{n} = 42$ and $\underline{n} = 21$. The species evolved as a result of the hybridization between an allotetraploid (4X) AABB ancestor and a wild diploid grass type containing DD. The tetraploid resulted from a hybrid between two diploid (2X) species with genome compositions of AA and BB respectively. Included in the tetraploid group is durum wheat used to produce macaroni and spaghetti products. The ancestors that contributed the A and D genomes have been determined cytologically to be primitive wheat, *Triticum monococcum,* and a wild grass, *Triticum tauschii* (formerly called *Aegilops squarrosa*). The origin of the B genome is still to be determined. The spikes of proposed bread wheat ancestors, along with normal bread wheat, are shown in Figure 5.1.

A serious cytological problem arises in interspecific crosses. If the chromosomes in the A genome are unlike those in the B genome, then the F_1 hybrid from the cross AA $\times$ BB will contain seven chromosomes from A and seven chromosomes from B, and no synapsis will occur in meiosis. Unpaired chromosomes in meiosis are called univalents, pairs are bivalents, associations of three are trivalents, and so on. Since bivalent formation is the basis of equal chromosome distribution and functional gamete production, it is necessary for the AB F_1 to double in chromosome number. This probably happened through unreduced gametes, and the allotetraploid AABB resulted. Meiosis could now proceed in a much more orderly fashion because of 14 pairs of chromosomes.

Figure 5.1. Hexaploid AABBDD bread wheat (single spike far right) and two spikes of each proposed diploid parent. Far left *Triticum monococcum* (AA); center *Aegilops speltoides* (possible BB); right: *Triticum tauschii* (DD). (Courtesy D. G. Wells, South Dakota State Univ.)

The same general course of action was followed in the hybridization of AABB × DD. The raw (undoubled) hybrid had 21 univalents. Doubling the chromosome number in the F_1 or producing unreduced gametes resulted in 21 bivalents and offered the opportunity for more nearly normal meiosis. The development of this polyploid system based on chromosome doubling in the F_1 is shown in Figure 5.2.

Combining two species represents a wide divergence from the normal genetic situation and is accompanied by extensive meiotic abnormality even after chromosome doubling. A long evolutionary process has taken place to produce the high degree of meiotic stability found in modern tetraploid and hexaploid wheat. In fact, if we did not know the cytological evolution of the species, there

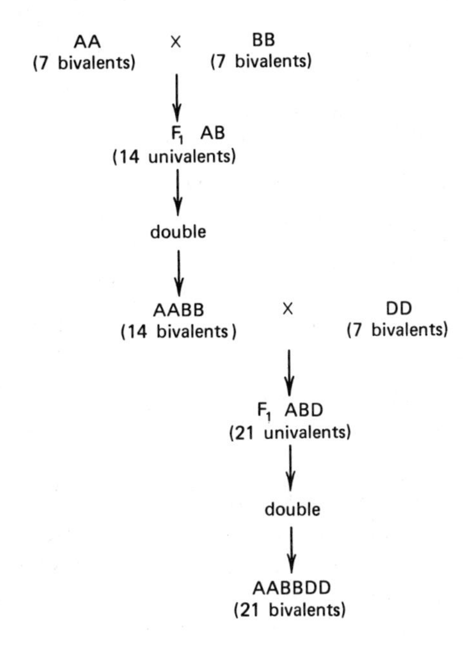

Figure 5.2. Genome composition and evolution of wheat based on chromosome doubling in the F_1 hybrid.

would be no way to separate the genomes and these plants would appear to be diploids. The term "amphidiploid" is used to describe those polyploids that are made up of different genomes but act like diploids in their chromosome pairing and movement.

The ancestral determination of polyploids is made through a number of evaluations including morphological similarities. The most convincing data, however, normally come from artificially created hybrids between the polyploid and the suggested ancestral species. If, in the hybrid F_1, pairing takes place between some chromosomes of the polyploid and those of the proposed donor, it is likely that the donor in question did provide a genome in the evolution of the polyploid. If no pairing occurs, the suggested donor is probably not an ancestral parent. Even the pairing measurements leave room for difficulty in interpretation, especially if a long period of chromosome evolution and morphological change has occurred since the original formation of the polyploid.

As an additional complication, chromosome pairing is, in itself, under genetic control. In the case of wheat, a gene on chromosome 5 in the B genome regulates pairing so that only homologous chromosomes will synapse. Removal of the 5B gene results in some pairing between nonhomologous chromosomes. Genetic control of pairing, while complicating evolutionary research, will be shown as very valuable in breeding through chromosome manipulations in Chapter 18.

Many polyploids contain genomes that have some chromosomes with a degree of similarity to those in other genomes. This condition, called segmental allopolyploidy, is usually very complex to interpret with respect to the origin of chromosomes because of some degree of pairing between genomes. The intergenomic pairing comes about as the result of chromosomes with similar segments in different genomes. Chromosomes with partial homology are termed homoeologous. Stebbins (10) describes the various kinds of polyploidy and the associated terminology in detail.

POLYPLOIDS AND GENETIC RATIOS

Polyploidy can produce interesting changes in genetic ratios and data interpretation. Since polyploidy involves multiple sets of chromosomes, it is logical to assume the possibility of a locus being represented on a multiple basis. This has been confirmed in the case of segmental allopolyploids.

In autopolyploids the segregation patterns are developed around the concept that identical chromosomes tend to synapse in meiosis I, which in turn governs the pattern of chromosome movement to the poles. For example, in an autotriploid the anaphase I chromosomes are usually distributed in a random manner so that one chromosome will go to one pole and two to the other. As mentioned earlier, the net result is the formation of gametes that contain varying numbers of each chromosome. The situation is further compounded by the fact that gametes with other than the full genome or some multiple thereof are often inviable, especially as pollen. An example of the segregation of one chromosome from a triploid individual is given in Figure 5.3. Remember, each locus is represented three times. We will assume a locus that has a domi-

♀ \ ♂	2*A*	1*a*
1*AA*	2*AAA*	1*AAa*
2*Aa*	4*AAa*	2*Aaa*
2*A*	4*AA*	2*Aa*
1*a*	2*Aa*	1*aa*

Segregation ratio: 2*AAA* : 5*AAa* : 2*Aaa* : 4*AA* : 4*Aa* : 1*aa*

Figure 5.3. An autotriploid segregation pattern from the cross ♀ *AAa* × ♂ *AAa*, assuming that only normal chromosome numbers are transmitted through the male.

nant *A* and recessive *a* allelic forms. The cross of ♀ *AAa* × ♂ *AAa* will be used to illustrate the segregation. If we assume that only those gametes that contain at least one allele are viable, then the random distribution of chromosomes will provide gametes in the ratio of 2*Aa* : 1*AA* : 2*A* : 1*a*. These can be used on the female side of the cross but the male will have only normal chromosome complements with a gamete ratio of 2*A* : 1*a*. The resulting progeny ratio is 2*AAA* (triplex) : 5*AAa* (duplex) : 2*Aaa* (simplex) : 4*AA* : 4*Aa* : 1*aa*. The genotype *aaa* (nulliplex) is not produced by this cross. The phenotypic ratio will depend on the type of gene action present. For example, if this is a situation where *A* is completely dominant, then the phenotype ratio would be 17 : 1. Students interested in studying these and other autopolyploid segregation patterns are referred to Burnham (2) and Allard (1) for additional information. Also, Busbice et al. (3) have provided an extensive summarization of research data on the autopolyploid condition and inheritance patterns in alfalfa.

Interesting genetic situations can also occur in segmental allopolyploids where each genome has probably evolved from some single original ancestor and thus contains some common genetic information. Consider an amphidiploid such as hexaploid wheat with three different genomes (A, B, and D) in the diploid condition. Sears (8) was able to show that the phenotypic effect of having one chromosome present four times could compensate for the complete lack of a specified chromosome in another genome, meaning that there was duplicated genetic information among the genomes. This in turn suggests that the chromosomes in each genome evolved from a common ancestor. As a result, the genetic segregation patterns in allopolyploids may deviate from that of normal diploids in much the same manner as autopolyploids do. Inheritance studies are often difficult to conduct in polyploids simply because of the partial duplication of genetic information and the compensating effects between chromosomes. Burnham (2) gives a thorough review of this topic.

ANEUPLOIDY

Aneuploidy is a variation in chromosome number that represents the loss or gain of one or a few chromosomes, but not an entire genome. This condition has apparently not played as significant a role in plant evolution as euploidy, but it does have important applications for the plant breeder and geneticist.

A chromosome present in the normal pair is the disomic or $2\underline{n}$ condition. If one chromosome is missing ($2\underline{n} - 1$), the situation is termed monosomic. A nullisomic ($2\underline{n} - 2$) is missing both chromosomes of a pair. If one chromosome of each of two pairs is missing, the individual is a double monosomic ($2\underline{n} - 1 - 1$). Other terminology for missing chromosomes can easily be extrapolated from these examples.

When an extra chromosome is present, the term "trisomic" ($2\underline{n} + 1$) is used. If two extra chromosomes of a pair are present, the individual is tetrasomic ($2\underline{n} + 2$). A double trisomic would have an extra chromosome for each of two pairs ($2\underline{n} + 1 + 1$). Various combinations of additional chromosomes may occur, but the cases listed are the most common.

Aneuploids can arise spontaneously as a result of gametes that have received less than the normal number of chromosomes. Partially unreduced gametes can be produced if nondisjunction (lack of separation) takes place in one or a few chromosome pairs in anaphase I. The cause of partial nondisjunction is not known but it can occur naturally with a low frequency. The resulting gametes will either have extra chromosomes or be missing some. Aneuploids may also be generated by gamete formation in such plants as triploids where chromosome pairing and movement is abnormal. Also, haploid plants can be pollinated with normal pollen and the results will occasionally be aneuploid.

Aneuploidy has been extremely valuable in locating genes on specific chromosomes and in understanding gene functions. Extensive studies have been carried out on several plant species including wheat, tomato, barley, rice, cotton, cress, tobacco, and snapdragon. Literature reviews on aneuploid investigations are presented in the *Handbook of Genetics,* Volume II (5).

We will consider only one of the many possible uses of aneuploids in genetic studies. In this example we identify the chromosome carrying a particular locus. Suppose we have a group of disomic plants with a homozygous dominant phenotype controlled by one gene pair (*AA*). In addition, nullisomic genetic stocks for all chromosome pairs are available. Furthermore, the nullisomic lines are all recessive for the character expression. We must only cross our dominant plants with each of the nullisomic lines and observe the segregation in the F_2 to locate which chromosome is carrying the *A* locus. This cross is diagramed in Figure 5.4. Assume that all functional male gametes from monosomic F_1's will have the full chromosome compliment. All populations will segregate in a normal 3:1 manner except for the critical F_2 family in which all individuals show the dominant expression. In this family the chromosome carrying the gene in question was contributed by the disomic parent only, as the nullisomic parent was missing both chromosomes of the pair, resulting in lack of segregation for the character in this population. The chromosome carrying the *A* locus has now been identified. While, in our example, we have considered only the use of nullisomics, many different types of aneuploids, including monosomics, trisomics, and others, have been valuable in chromosome analyses by using the abnormal segregation principle.

Aneuploids can also be used to determine the effects of chromosome and gene dosages. For example, Rick et al. (7) described a complete collection of trisomics in tomato and showed that each

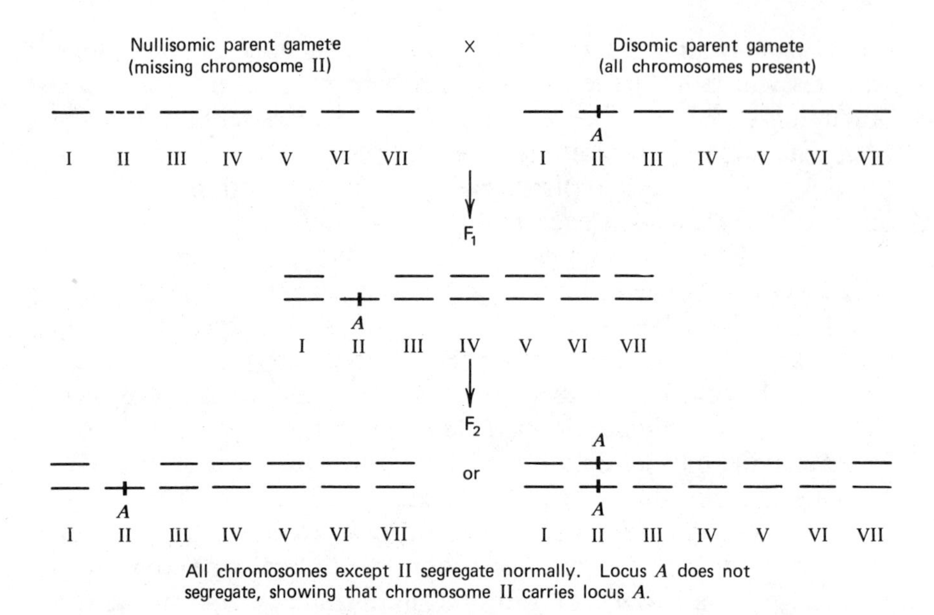

Figure 5.4. The use of nullisomic lines to identify the chromosome containing a particular locus. Assume that the deficiency is not transmitted through the male.

trisomic could be identified by a highly distinctive phenotype. The phenotype variation resulted from three copies of the gene being present, instead of the usual two. Many other studies of this type are mentioned in the *Handbook of Genetics* review articles (5). The accurate evaluation of genetic studies involving aneuploids calls for a high degree of competence in microscopic techniques associated with chromosome number and morphological analysis. Also, extensive effort goes into the breeding and development of the various types of aneuploid stocks so that the data resulting from their use can be interpreted properly. Much sophisticated research in plants has been conducted around aneuploid analyses.

SUMMARY AND COMMENTS

Variations can occur in chromosome numbers through euploidy and aneuploidy. Polyploidy gives a genetic mechanism for combining sets of chromosomes either within or between species. Aneuploidy provides a powerful tool for the study of inheritance systems and genetic mechanisms.

Researchers in the science of chromosome investigations can piece together the evolutionary aspects and relationships of the bearers of genetic information. Some of the most elegant studies in biological science have been conducted by these highly skilled and patient chromosome detectives. Their contributions have been very valuable to plant breeders by providing more information about the nature and source of genetic variation.

REFERENCES

1. Allard, R. W. 1960. *Principles of plant breeding.* Wiley, New York.

2. Burnham, C. R. 1962. *Discussions in cytogenetics.* Burgess, Minneapolis, Minn.

3. Busbice, T. H., R. R. Hill, and H. L. Carnahan. 1972. Genetics and breeding procedures, pp. 283-318. In C. H. Hansen (ed.), *Alfalfa science and technology.* Am. Soc. Agron., Madison, Wisc.

4. Darlington, C. D., and A. P. Wylie. 1955. *Chromosome atlas of flowering plants.* George Allen and Unwin, Ltd. London.

5. *Handbook of genetics,* Vol. 2. 1974. R. C. King (ed.). Plenum Press, New York.

6. Löve, Askell, and D. Löve. 1961. *Chromosome numbers of central and northwest European plant species.* Almquist and Wiksell, Stockholm.

7. Rick, C. M., W. H. Dempsey, and G. S. Kusch. 1964. Further studies on the primary trisomics of tomato. *Can. J. Genet. Cytol.* 6:93-108.

8. Sears, E. R. 1965. Nullisomic-tetrasomic combinations in hexaploid wheats, pp. 29-45. In R. Riley and K. R. Lewis (eds.), *Chromosome manipulations and plant genetics. Supp. to Hered.* 20.

9. ———. 1974. The wheats and their relatives, pp. 59-91. In R. C. King (ed.), *Handbook of genetics,* Vol. 2, Plenum Press, New York.

10. Stebbins, G. L. 1949. Types of polyploids: their classification and significance. *Adv. in Gen.* 1:403-409.

QUESTIONS

1. Suppose a species has a somatic chromosome number of 36. When crossed with another species that has a somatic chromosome number of 12, 6 bivalents appear in the F_1.

a. How many univalents does the F_1 have? **Answer:** 12.

b. Is the second species a genome doner of the first species? **Answer:** Yes.

c. What is the ploidy level of the first species? **Answer:** Hexaploid.

d. Give the chromosome formulation for each species. **Answer:** First species: $2\underline{n} = 6X = 36$
Second species: $2\underline{n} = 2X = 12$

2. Assume in an autotriploid that *R* is resistant and completely dominant over susceptible *r*. What is the result of the cross ♀ *RRr* × ♂ *Rrr*? **Answer:** 1*RRR* : 4*RRr* : 4*Rrr* : 2*RR* : 5*Rr* : 2*rr* or 8 resistant : 1 susceptible

3. Assume height is governed by a single locus, with tall (*T*) being completely dominant over short (*t*). A nullisomic individual that is short is crossed as a female with a disomic tall male.

a. What will the F_2 segregation be if the height locus is on the nullisomic chromosome? **Answer:** All tall.

b. If the F_1 monosomic for the chromosome carrying the height allele is crossed as a male with a disomic short plant, what will the progeny be? **Answer:** All tall (since chromosome abnormalities are normally not passed through the male).

c. If the progeny in 3b are intercrossed, what is the resulting ratio? **Answer:** 3 tall : 1 short.

6

PLANT REPRODUCTION

An almost overwhelming array of reproduction mechanisms exists in the plant kingdom. The systems run the gamut from extremely elaborate flowering displays with sophisticated external vectors to simple vegetative cuttings. The plant breeder must be aware of the alternatives in order to apply correct breeding and improvement techniques. Knowledge of the physical mechanisms and genetic implications associated with various reproductive systems are imperative in planning and conducting a breeding program. In this chapter we consider mechanisms and the genetic consequences of sexual and asexual reproduction.

SEXUAL REPRODUCTION

In sexual reproduction the offspring are produced by the union of male and female gametes. The discussion in this section is limited to true sexual combinations in higher plants. We will later take up cases where progeny are produced by what appear to be sexual but in reality are asexual processes.

Gamete Formation

All gametes involved in sexual reproduction are produced through meiosis. In male gamete formation (microsporogenesis), four haploid cells are produced from each complete meiotic division. These cells, called microspores, mature to form pollen grains in the anther. Prior to pollination the nucleus in the pollen divides mitotically to form two nuclei. One of these, the tube nucleus, will remain intact. The other will divide once more to form two gener-

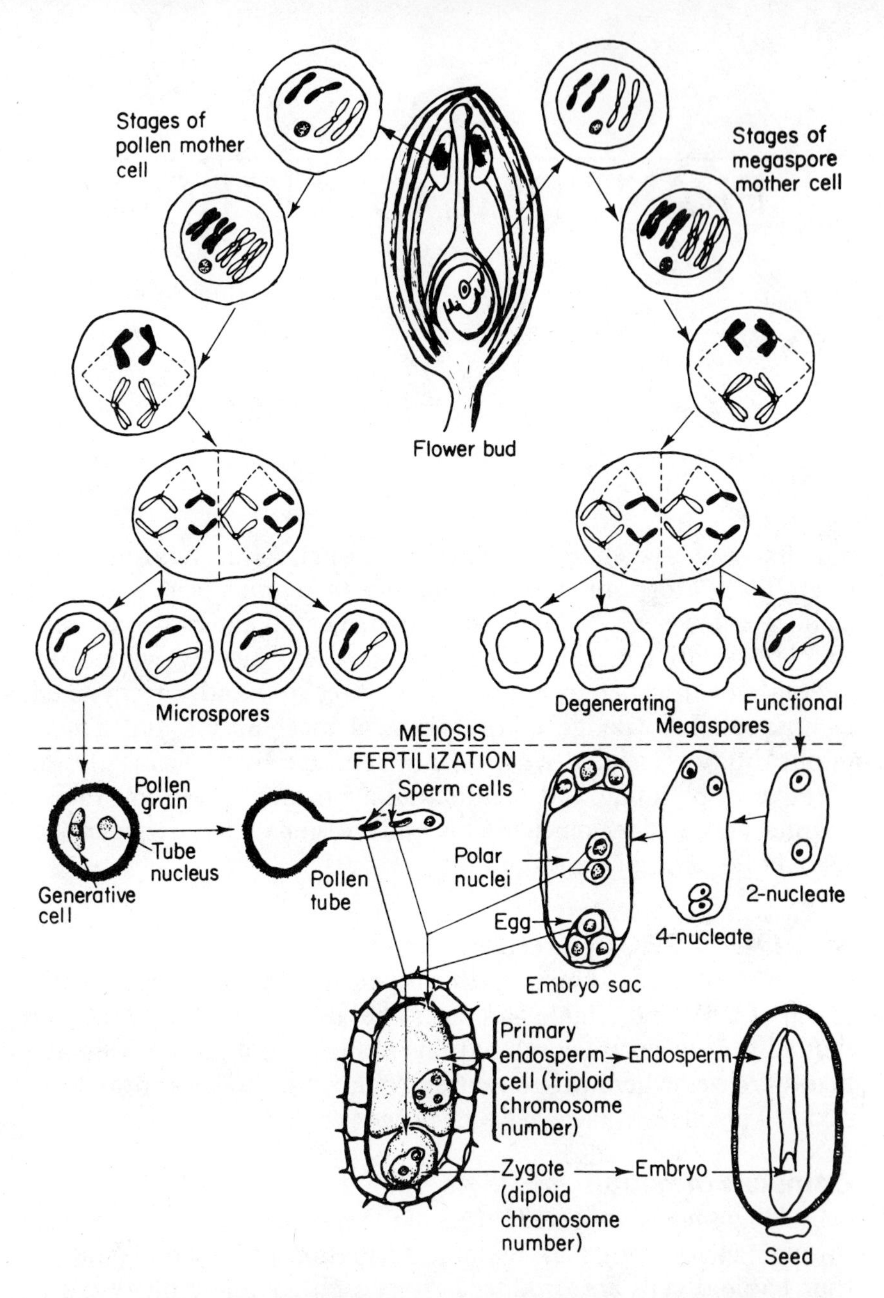

Figure 6.1. Sexual reproduction in angiosperms. (From Hudson T. Hartmann, Dale E. Kester, *Plant Propagation: Principles and Practices,* 3rd ed., 1975, p. 10. Reprinted by permission of Prentice-Hall, Inc., Englewood Cliffs, N. J.)

ative or sperm nuclei. The net result is a pollen grain containing three nuclei each with a 1n chromosome complement. See Figure 6.1.

Anthesis is the maturation of the anther accompanied by the extension of the filament, which is the structure supporting the anther. In many species, the elongation of the filament is the means by which the anther is extruded from the flower, making the pollen available to other plants. As the final point of maturity is reached, the anthers will dehisce (experience dehiscence) or rupture to release the mature pollen load.

In female gamete formation (megasporogenesis), three of the four haploid cells produced by meiosis in the megaspore mother cell degenerate. The remaining cell enlarges to become the embryo sac. There are several variations of the subsequent steps so only the general case is presented here. Three successive mitotic divisions occur so that eight 1n nuclei are produced from the original nucleus. The nuclei migrate into the positions shown in Figure 6.1. The egg and two synerged cells lie at the end of the ovary near the opening or micropyle. Three antipodal cells are located at the end of the ovary opposite the micropyle. Two polar nuclei remain in the center of the embryo sac.

Fertilization

Pollination, the arrival of the pollen on the stigma, initiates the processes leading to fertilization. The pollen develops a tube that grows through the stylar tissue toward the micropyle. During this phase the tube nucleus moves ahead of the two sperm nuclei. See Figure 6.2*a*. When the pollen tube reaches the micropyle, the tube nucleus degenerates and the two sperm nuclei enter the embryo sac.

At the moment of fertilization (Figure 6.2*b*) one sperm nucleus combines with the egg to form the zygote and the other unites with the two polar nuclei to produce the endosperm. This double fertilization process results in a 2n embryo and a 3n endosperm. About 85 percent of the angiosperms follow this procedure. The remainder experience double fertilization but contain variations principally in the number of nuclei present in the embyro sac at the time of fertilization.

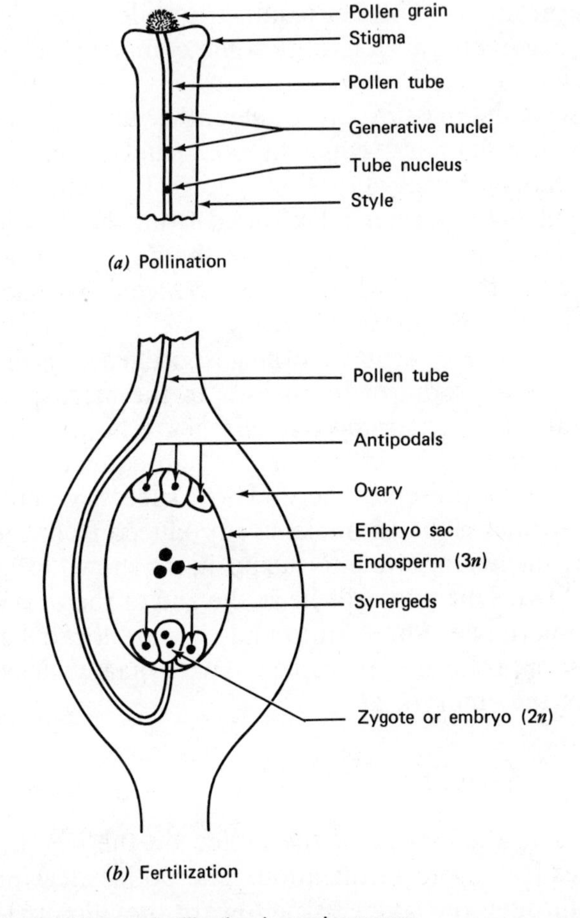

Figure 6.2. Pollination and fertilization in angiosperms.

Male gametes, as pollen grains, are generally produced in far greater numbers than can ever be used. This is amply demonstrated by the abundance of pollen in a corn field at tasseling time or in a meadow of flowers in full bloom. Overabundant production is one method of counteracting high pollen mortality. While many pollen grains can land on a single stigma and possibly even initiate tube growth, the prize goes to the first male gamete completing the journey and fertilizing the egg. This provides a highly effective screening mechanism against gametes that are noncompetitive because of genetic defects or are incompatible with the female tissue. Overabundance of pollen production has undoubtedly played a significant role in the evolutionary process.

The endosperm provides a nutrient support system for the young embryo during its early stages of growth but no genetic input into the next generation. This is dictated entirely by the zygote. The endosperm does play a significant role in the economic aspects of many plants. For example, crops such as wheat, barley, oats, rice, and corn are produced for their grain. Productivity depends on the ability of the plant to provide a large number of well-developed endosperms with correct physical and chemical properties. Because of the double fertilization process, the female will always contribute two alleles to the endosperm genotype and the male one. Thus the cross of ♀ *AA* × ♂ *aa* will have an endosperm of *AAa*. The reciprocal cross of ♀ *aa* × ♂ *AA* will possess an *aaA* endosperm. Two heterozygotes crossed can produce all possible combinations, and an allele at one locus can be present zero, one, two, or three times.

Endosperm size and development are affected genetically in two ways. First, genetic information contained in the endosperm is reflected in its ultimate phenotype. Good examples of this have been worked out in corn. In some genotypes the endosperm may have a dotted color pattern caused by a recessive allele a_1. Rhoades (14) demonstrated that the number of dots in each kernel increased with the number of a_1 alleles present in the endosperm. In another corn endosperm study, Mangelsdorf and Fraps (8) showed a direct positive relationship between the number of dominant yellow pigmentation alleles (*Y*) and the important character of vitamin A content. As the number of *Y* alleles increased, the yellow pigmentation intensity and subsequent vitamin A content also increased.

Second, the development of the endosperm is controlled by the genetic composition of the plant that is reflected in the way the plant functions and produces. A genotype that does not perform well under stress, for example, will produce poorly developed seed that is, in large part, endosperm.

In dicotyledonous plants such as peas that are produced for their grain, the endosperm plays a small role and the cotyledons become the economically important structures. These are part of the 2n zygote and follow normal segregation patterns. When Mendel studied seed color in peas he was actually looking at the cotyledon color. Thus his proposed inheritance pattern in the seed was not confused by the 3n endosperm condition.

Natural Pollination Systems

Pollination systems are normally divided into two broad classifications. Self-pollination occurs when the pollen from a plant unites with the stigma from either the same flower or a flower on the same plant. Cross-pollination takes place when the pollen from one plant combines with the stigma from a different plant. Open-pollination means that plants are free to pollinate with themselves or other plants.

Equality is often assumed between the terms pollination and fertilization. However, the student is cautioned to remember that pollination does not automatically insure fertilization. Self- and cross-pollination can occur without the subsequent union of gametes. Self- and cross-fertilization (outcrossing) correctly describe zygote formation.

Pollination and Fertilization Control Mechanisms

A wide range of pollination and fertilization control systems exist in nature. Falgri and van der Pijl (5) and Percival (12) provide detailed information on pollination and fertilization evolution and function. Our considerations include some of the more common mechanisms encountered in plant breeding programs.

In the plants that are readily self-fertilized, ample opportunity exists for pollen to reach the stigma of the same flower, or the flowers of the same plant. Also, the pollen is completely functional in the initiation and growth of the tube through the style, and fertilization proceeds normally. A high degree of self-fertilization does not, however, automatically eliminate the possibility of cross-fertilization. Table 6.1 lists some examples of plants that are highly self-fertilized. In many of these cases cross-fertilization can occur with a low frequency. Barley, for example, is highly self-fertilized but may outcross up to 5 percent. The degree of outcrossing depends on both the genotype and environmental factors

Table 6.1. Examples of Commonly Self-Fertilized Plants

Apricot	Peas
Barley	Potato
Beans	Rice
Cotton	Sorghum
Flax	Soybeans
Lettuce	Sweet clover
Nectarine	Tobacco
Oats	Tomato
Okra	Vetch
Peach	Wheat
Peanut	

such as temperature, wind, and moisture. Long cool weather periods during anthesis in the self-fertilized cereals promote pollen longevity and an increased duration of stigma receptivity. Hot, dry, windy conditions greatly shorten pollen life and reduce cross-fertilization.

The general mechanism to insure self-fertilization involves self-pollination before the stigma is exposed to any foreign pollen. Breeders will do this by mechanical means such as bagging the flowers before they reach sexual maturity. In nature, a perfect flower with the stamens and pistil in the same infloresence is an aid, but no guarantee to self-pollination. Cleistogamy, where the flower never opens, is the most foolproof method of insuring self-fertilization. A list of cleistogamic species is provided by Uphof (15). Variations in the expression and degree of cleistogamy are produced by different genotypes and environments. Erickson (4) reported differing degrees of cleistogamy with different varieties of soybeans and also found that environment affected the degree of cleistogamy in each variety.

A less severe but highly effective mechanism to promote self-fertilization is provided in those plants where flowers open (chasogomy) but where pollination and fertilization take place prior to the opening. Wheat, barley, oats, and rice are examples of species with this system. They will experience a higher degree of outcrossing than cleistogamic types but will still be primarily self-fertilized because of pollination occurring early in the flowering process. Often, we describe fields of these crops as "blooming" or "flowering." At the visible bloom stage, the anthers that have extruded from the florets and characterize the flowering appearance are indications of an already completed fertilization event. The absolute level of outcrossing in chasogomous species is dependent on both genotype and environment.

As we move along the scale in the direction of cross-fertilization we find a number of species that may experience outcrossing up to 50 percent of the time. Included in this group are cotton, sorghum, and sudangrass. While they are still quite highly self-fertilized, they tend to have more stigma exposure and an occasional slight mismatch of pollen and stigma maturity. Both these conditions aid in promoting cross-pollination. Wind, in the case of sorghum and sudangrass, and insects in cotton are the principal pollen vectors.

This brings us to the multitude of species that are primarily cross-fertilized. A partial listing of some plant species experiencing 50 percent or more cross-fertilization is given in Table 6.2.

Monoecy, in which the male and female reproductive organs are on differing parts of the same plant, promotes cross-pollination. Corn, where the tassel produces the male gametes and the ear the female, is a common example. Monoecy does not, in itself, insure cross-pollination. Corn can be self-pollinated and self-fertilized quite easily with standard isolation and flower manipulation techniques. However, the tremendous pollen load in a field of corn coupled with the differences in time of maturity between the anthers in the tassel and the silks or stigmas result in a very high level of cross-pollination.

Protandry, the maturation of the stigma prior to the pollen, and protogyny, the maturation of the pollen prior to the stigma, both lend themselves to cross-fertilization. A great deal of variation exists within these mechanisms. For example, some species may have some slight overlapping of maturation between the male and the female in hermaphroditic (both sexes present) flowers. This could be interpreted as a safety factor in providing seed set

Table 6.2. Examples of Commonly Cross-Fertilized Plants

Alfalfa	Hemp[a]
Almond	Hops[a]
Alsike clover	Olive
Apple	Onion
Asparagus[a]	Pear
Avocado	Pecan
Banana	Plum
Beet	Pumpkin
Birds-foot trefoil	Radish
Blackberry	Raspberry
Blueberry	Red clover
Broccoli	Rye
Cabbage	Rye grass
Carrot	Smooth bromegrass
Celery	Spinach[a]
Cherry	Strawberry
Corn	Sunflower
Cucumber	Sweet clover
Date[a]	Walnut
Fig	Watermelon
Grapes	

[a]Dioecious.

and reproduction by self-pollination if normal cross-pollination has not occurred. As in many other cross-pollination promotion mechanisms, the presence of protandry and protogyny does not entirely preclude self-fertilization.

The most dramatic mechanism to insure cross-pollination is that of dioecy. Here separate male and female plants occur, and reproduction by gamete combination requires cross-fertilization. Hops, hemp, asparagus, spinach, and the date palm are examples of dioecious plants. The genetic control of sex determination in dioecious plants, first described in the early 1900s, becomes extremely important in plant breeding strategies, particularly in hybrid programs. Dioecy, for example, is apparently governed by several loci responsible for either stamen or carpel suppression. These loci are in turn controlled by a master locus where the alleles *M* or *m* can occur. If the plant is *Mm* it is a male (staminate), while *mm* results in a female (pistillate) plant. Note that an *MM* genotype is impossible. The efficient operation of sexual dimorphism in several species has resulted in the establishment of linkage relationships within each group of sex suppression loci. The linkage is protected against recombination through crossing over by the evolution of hetermorphic sex chromosomes similar to those found in some animal species. A summary of dioecy genetics has been provided by Westergaard (16).

Another interesting system of genetically controlled sex expression is provided by the group of plants that includes the cucumber. When the flower of the cucumber is initiated, it has the initial stages of both stamen and ovary. From here on, a wide array of possibilities exist. In many varieties monoecism occurs with the first flowers being predominantly male and the later ones more female. However, some varieties are andromonoecious—meaning that some flowers are male while others are hermaphroditic. Gynoecious genotypes produce only female flowers. Additional variations in sex expression are also possible. The implications of their potential utilization in hybrid crops are discussed in Chapter 16. While these and other variations on sex expression are under genetic control, they can also be altered by chemical treatment and natural environmental factors. Frankel and Galun (6) provide an excellent discussion and review of recent literature on this subject.

The final item in our consideration of cross-fertilization promotion mechanisms is self-incompatibility where pollen tube growth is related to specific male and female genetic combinations. Pandey (9) indicates that self-incompatibility has been found in 78 angiosperm families and occurs in every major phylogenetic line. It offers special production challenges in several important crop groups such as the legumes, fruits, and vegetables. Because of self-incompatibility, for example, more than one genotype may have to be available before fruit production can be achieved in an apple or pear orchard. From the plant breeding standpoint, self-incompatibility can seriously limit the possible genetic combinations produced through crossing.

For our purposes a short generalized discussion on incompatibility mechanisms is sufficient. The student should be aware, however, that many modifications of these generalizations occur. Pandey (9,10,11) has provided excellent review articles on the evolution and description of self-incompatibility.

The operation of self-incompatibility centers around the idea that pollen that is incompatible with the stylar tissue will either not germinate on the style or will produce very slow pollen tube growth. In this way fertilization is limited to compatible pollen that germinates and produces normal pollen tube growth, even though both kinds of pollen may land on the stigma simultaneously.

Incompatibility, or compatibility, is dictated by a genetic system operating in both the male and the female. The functionality of the system is based on a physiological interaction between the pollen ($1\underline{n}$) and the style ($2\underline{n}$). A multiple allele series, designated *S*, governs the system. Traditionally, a single locus has been proposed to control the mechanism, but Pandey (11) indicates the possible involvement of two loci in some species. Our discussion will assume a single locus in which an allele series designated S_1, S_2, S_3 . . . has been generated through mutations. The number of alleles varies with the species.

Incompatibility can be expressed in one of two systems, either gametophytic or sporophytic. In the gametophytic system, incompatibility results when a pollen grain and the stigma have an allele in common. For example, the cross ♀ S_1S_2 × ♂ S_1S_2 would be incompatible because the pollen would carry either S_1 or S_2, both of which are common with the stylar tissue. However, the cross ♀ S_1S_2 × ♂ S_1S_3 would be compatible and produce S_1S_3 and S_2S_3 offspring because the male gamete carrying S_3 would function normally. The reciprocal of this cross would also be compatible and produce S_1S_2 and S_2S_3 progeny. Theoretically, in the gametophytic system a homozygous S allele conditon is not possible through crossing.

The sporophytic system contains a form of dominance in which S_1 is dominant over all other alleles, S_2 is dominant over all but S_1, and so on. In microsporogenesis all pollen, regardless of genotype retains the phenotypic response of the dominant allele in the male diploid tissue. For example, an S_1S_2 male would produce pollen with S_1 phenotype, even though some were of the S_2 genotype. There is no dominance expressed on the female side and the female functions in exactly the same manner as in the gametophytic system. In the sporophytic system the cross ♀ S_1S_2 × ♂ S_1S_3 is incompatible because the dominance effect on the male side dictates that both S_1 and S_3 pollen have the phenotype of S_1. Since S_1 is incompatible with the S_1S_2 stylar tissue, no fertilization occurs. The reciprocal of this cross is also incompatible. The sporophytic system, in contrast to the gametophytic system, does allow for some homozygosity of the S alleles. For example, the cross ♀ S_2S_3 × ♂ S_1S_2 would produce progeny of S_1S_2, S_1S_3, S_2S_2, and S_2S_3.

The difference between the sporophytic and the gametophytic mechanisms is thought to be based on the timing of biochemical substance development associated with the incompatibility response of the pollen. This may occur very late during pollen development in the gametophytic system and very early in the sporophytic system.

From this discussion, it is evident that self-incompatibility strongly promotes cross-fertilization in many species. Like all other mechanisms, however, it is not completely foolproof. Considerable success has been achieved in finding, or producing artificially, mutations at the S locus that will produce self-fertility in normally self-incompatible species. In other cases, where pollen tubes are very slow growing, self-fertilization can be obtained if all foreign pollen is excluded. Occasionally, the removal of part of the stylar tissue or mechanical treatments, such as heat and cold, applied to the style will result in some self-fertility. Thus, self-incompatibility can be considered to be a mechanism that encourages the mating of unlike genotypes but allows a certain amount of flexibility and "leakage" in genetic recombinations.

Determining the Fertilization System

A knowledge of the fertility system is important in approaching the genetic improvement of any species because this mechanism governs the ease with which new genetic combinations are made. Is the plant highly self-fertile or is it almost always cross-fertilized? If cross-fertilization takes place regularly, what mechanisms are responsible, are there possibilities for self-fertilization, and, if so, what are the genetic consequences? These are common questions confronting breeders in programs where little information is available about the plant.

The amount of selfing and outcrossing in a highly selfed crop like barley or wheat is not difficult to determine. The utilization of simple genetic markers such as color have been effective in the studies. For example, a variety carrying a homozygous dominant allele for red glumes is planted beside a variety homozygous for white glumes. Seed is produced and harvested from the white glumed variety. The seed is planted and the number of red glumed plants present in the population are an indication of the amount of cross-fertilization between female gametes from the white variety and male gametes from the red variety. We must remember, however, that this test has only evaluated the outcrossing between these two parent plants under a specific environment. Other environments and other varieties may produce different results.

Evaluating species with higher natural levels of outcrossing may not be so simple and generally some degree of spatial or mechanical isolation is required to determine the amount of selfing or outcrossing. In spatial isolation, plants are prohibited from receiving foreign pollen simply because of distance. This may be impractical in many cases because of a common presence of the species in a large geographic area.

The most widely used alternative is to provide some form of mechanical isolation such as bagging or caging. Several problems are encountered in these programs. With bags, for example, environmental conditions such as heat, light, and relative humidity may be changed so that the data are a reflection of the isolation mechanism treatment as well as the fertilization system itself. Also, if pollen vectors such as insects are required in the natural system, isolations should be designed to either include the insects or provide some mechanical alternative of pollen transport. Also, with self-incompatibility selfing may be achieved simply because no foreign pollen is available.

Even when the degree of selfing or crossing has been established, the question of mechanisms controlling the fertilization still remains and these mechanisms are generally the subject of many innovative investigations often requiring a team of breeders and geneticists. The extensive studies of the self-incompatibility system are a good example. In some cases the breeder may have to rely on available information and design the breeding program around the systems as observed, without entirely understanding the complete mechanism.

Artificial Pollination Systems

The heart of any breeding program lies in the ability to produce new genetic combinations and the subsequent selection of potentially better types. To accomplish this, male and female gametes from desired parental genotypes must be brought together (crossed). The production of a cross is often considered the most important step in the breeding program. Here, a distinction should be made between the mechanical operation of crossing, and the production of progeny from specific parental combinations. Making a cross may take only a few moments of physical manipulation. Designing the parental combinations, on the other hand, generally requires all the background knowledge, training, and intuition that a breeder can muster. This discussion is limited to the mechanics of making a cross. The question of parent selection will be covered in chapters on breeding programs.

It would be impossible to cover every crossing system within each crop in detail because of the variability in flowering mechanisms within and among species. This diversity dictates that a large number of crossing approaches and innovations be used. A few main points will be established, and the student is encouraged to study specific techniques with the aid of suggested readings, laboratory exercises, and in cooperation with professional breeders. Poehlman (13) has provided details of crossing mechanics for most of the agronomic crops. An extensive summary of crossing mechanisms in many crop species is presented in the publication *Hybridization of Crop Plants* (7). Janick and Moore (1) have compiled a good array of information for fruits and nuts. Barrett and Arisumi (3) also provide information on pollen collection and crossing techniques in fruit breeding. Many ornamental species have individual publications available in which crossing techniques are described. Most breeders are generally happy to share their information with interested students and may, in many cases, provide part-time jobs so students may try their hand at crossing and other breeding procedures.

The object in any crossing program is to bring the desired male and female gametes together. The term "nick" or "nicking" has been used to describe the successful matching of a male and female to produce a cross. It has also been used to indicate those parental combinations that produce a high proportion of desirable progeny. Either use is acceptable. Figures 6.3 to 6.8 provide examples of a few crossing techniques in several species.

In species with hermaphroditic flowers, the removal of anthers (emasculation) prior to anthesis may be required. This is particularly important in self-fertile plants but may not be as critical in species with a high degree of self-sterility. Emasculation techniques vary greatly, depending on the size of the anthers, the position within the flower, and the relative time of maturity between the anthers and the stigma. Common methods of anther removal include the use of tweezers, scissors, the thumbnail, a toothpick, or some type of suction apparatus. Good vision and a steady hand are common requirements in the emasculation of many small flowered species. In some cases, emasculation is accomplished by exposing the anthers to environmental stress such as hot water, which may result in pollen death. Monoecious and dioecious plants are quite easy to handle. Corn, for example, can be emasculated simply by breaking off the tassel.

Once the necessary precautions are taken to exclude unwanted pollen, the next step is to make the pollination. The timing of this operation depends on the presence of mature pollen at the time of stigma receptivity. The time from emasculation to pollination differs among species. In cereals, for example, pollination should be accomplished about one to five days after emasculation,

Figure 6.3. Corn crossing. (*a*) A tassel (male) undergoing dehiscence. (*b*) An ear (female) extruding the silks (stigmas). (*c*) The tassel is bagged to collect the pollen. (*d*) The silk is clipped from the young ear that is then bagged to exclude foreign pollen. (*e*) Pollination is carried out by placing a bag containing the pollen on the female that has extruded new silks. (*f*) The pollen bag is fastened on the ear and shaken vigorously. (*a, b, c, d,* and *f*—Courtesy W. A. Russell, Iowa State Univ., *e*—Courtesy D. G. Wells, South Dakota State Univ.)

(c)

(d)

Figure 6.3. (*Continued*)

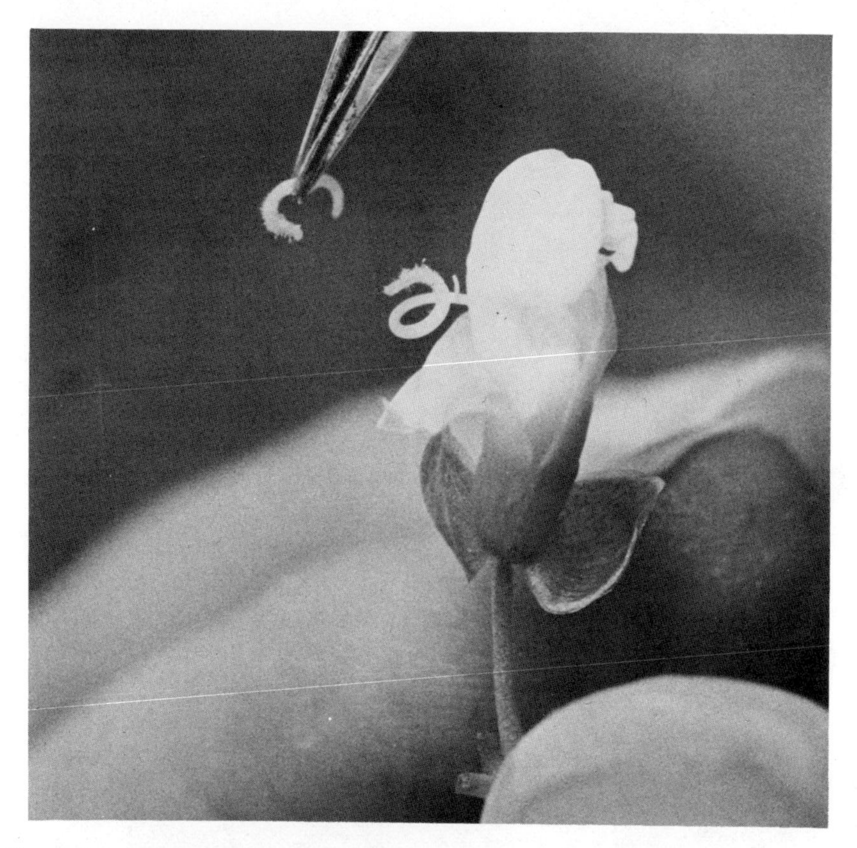

Figure 6.4. Bean crossing. The anthers are removed and the stigma is carefully brought out of the petal enclosure. Pollination is being accomplished by bringing a pollen-laden stigma from the male plant to the female. When the two stigmas are brushed together, the pollen is transferred to the emasculated female flower. (Courtesy D. P. Coyne and Carol Erwin, Univ. of Nebraska.)

while in some species emasculation and pollination can be carried out in the same operation.

Pollen can be applied in a variety of ways. Flowers that are undergoing dehiscence can be placed in close proximity to the female, bagged together, and periodically shaken to stimulate pollen movement. Pollen may be collected from desired males and manually applied to the stigma with a brush or tweezers. In cleistogamic species, stigmas from recently self-pollinated flowers of

Figure 6.5. Apple crossing. (*a*) Emasculation by picking the petals and anthers with the fingertips. (*b*) Following the collection of pollen in a dish, the females are pollinated by dipping a finger in the pollen and brushing the stigmas. (Courtesy R. C. Lamb, New York State Ag. Exp. Sta., Geneva.)

Figure 6.6. Cherry crossing. A large isolation cage is inverted over the tree. With the strong self incompatibility system, emasculation is not necessary. Bees are introduced into the cage along with bouquets of dehiscing male branches. Natural cross pollination is then allowed to take place. (Courtesy R. C. Lamb, New York State Ag. Exp. Sta., Geneva.)

desired males can be removed with their pollen load and brushed against the emasculated female. Whatever the technique, precautions must be taken to insure the application of the proper pollen to the desired stigma. This may require some sterilization of equipment between pollinations when using different males.

Pollen longevity is a serious consideration in crossing. Cereal breeders are usually limited to obtaining pollen from male plants

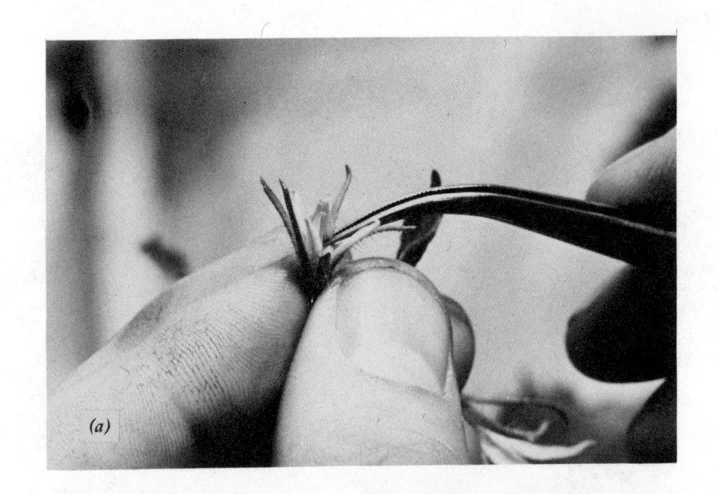
(a)

(b)

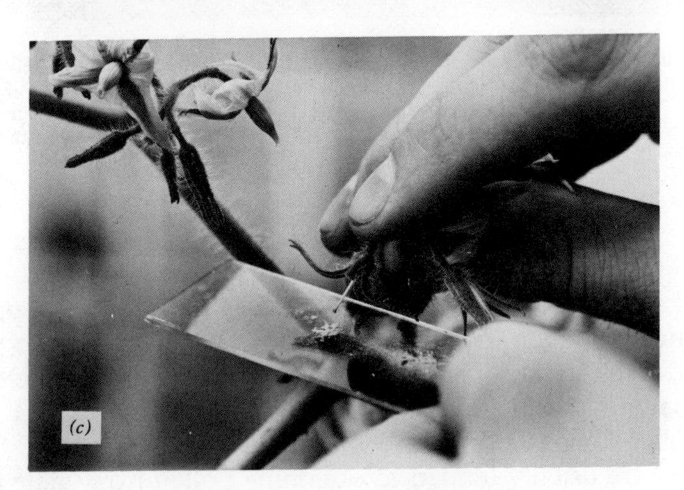
(c)

on the spot, as pollen life in many grasses is only a few minutes. Fruit breeders, on the other hand, may be able to collect pollen for some species and store it over a period of months or years while retaining usable levels of viability. Another common technique in tree species involves the storage of cuttings or branches with flowers in a dormant condition that are, at the proper time, environmentally stimulated to flower and dehisce for crossing with the females.

Some form of isolation such as bagging, both before and after pollination, may be necessary to exclude unwanted pollen. Care must be taken to maintain an environment that is not detrimental to fertilization and embryo development.

Recordkeeping is maintained throughout the crossing process. The amount of information necessary is a function of the requirements of the specific breeding program. Records may be as detailed as the identification of each parent in a specific cross, or as general as the knowledge of a group of parents used in operating some kind of mass crossing system. Almost without exception, however, some records are kept to inform the breeder about the proven productivity of particular parents to be considered in future crosses, and the genetic control of characters in segregating populations. Following the crossing procedure, a common practice is to attach a tag containing pertinent information to each female.

Breeders must know the environmental requirements of their plants so flowering will occur on a programmable basis. Both temperature and photoperiod play a major role in the flowering mechanisms of most species. These processes should be well understood to keep the nicking of desired parents, and crossing programs, on schedule. Crossing may be carried out either in the field or greenhouse. Many breeders find it convenient and efficient to conduct crossing programs in greenhouses during those seasons of the year when field work is at a minimum. This flexibility will

---

Figure 6.7. Tomato crossing. (*a*) Emasculation is done by tweezer removal of immature anthers. (*b*) Pollen is collected from dehiscent anthers by stimulating with a vibrating needle. (*c*) The stigma in the emasculated flower is pollinated by dipping the tip into the collected pollen. (Courtesy A. K. Stoner, USDA-SEA.)

(a)

(b)

(c)

Figure 6.8. Grass crossing. (*a*) The sexual parts of a grass flower. Each of the three stamens is composed of a rod-shaped anther and the supporting filament. The pistil is made up of two stigmas with many hairy extensions, stylar tissue, and an ovary. (*b*) Emasculation by tweezer removal of anthers from clipped florets. (*c*) Pollination with a spike from which dehiscent anthers are extruding. The inverted male is twirled vigorously and the pollen falls on receptive stigmas in the bagged spike. (*a, b*—Courtesy W. Dewey, Utah State Univ., *c*—Courtesy S. Campbell, Univ. of Florida.)

often depend on the species, as some do not lend themselves to artificial environments and confinement.

GENETIC IMPLICATIONS OF SELF- AND CROSS-FERTILIZATION

The genetic composition of resulting populations differs with self- and cross-fertilization. These differences are discussed in a general way here and will be applied to specific breeding programs in later chapters.

Self-fertilization

Populations in self-fertilized species tend to become highly homozygous. This can be easily demonstrated using two alleles at one locus. If a heterozygous plant *(Aa)* is selfed, a $\frac{1}{4}AA : \frac{1}{2}Aa : \frac{1}{4}aa$ ratio is produced. The homozygous dominant and recessive individuals are now fixed and will produce only homozygous dominant or recessive genotypes in the next generation. The heterozygotes, when selfed, will produce progeny in a $\frac{1}{4}AA : \frac{1}{2}Aa : \frac{1}{4}aa$ ratio and in effect reduce the proportion of heterozygotes in the total population by 50 percent. Thus, the population is being increased in homozygosity by 50 percent with each successive generation of selfing. The amount of heterozygosity at one locus in any generation can be expressed by the formula $(\frac{1}{2})^n$ where n is the number of segregating generations. For example, with one locus only one-eighth of the population is heterozygous in the F_4. If selfing without selection continues long enough the population will, for all practical purposes, be made up of 50 percent homozygous dominant and 50 percent homozygous recessive individuals.

Complexity is introduced into the system by adding additional loci. The number of completely homozygous individuals will depend on both the number of loci and the number of generations of selfing. The formula $[(2^m - 1)/2m]^n$ can be used to calculate the proportion of homozygous individuals in each generation. The term m represents the number of segregating generations, and n is the number of loci involved. For example, with a three locus system in the first segregating generation (F_2), one-eighth of the individuals will be homozygous at all three loci. This, of course, includes those that are homozygous dominant at all three loci, dominant at one and recessive at the other two, and so on. In the F_3 of this same population the homozygous genotypes now comprise 27/64 of the population, representing a considerable increase in the proportion of homozygous individuals with just one generation of selfing.

The number of population members that have various degrees of genotype homozygosity can also be calculated. An expansion of the binomial formula $[1 + (2^m - 1)]^n$ will provide this information. Again, m equals the number of segregating generations and n is the number of loci. In this expansion the exponent of the first calculated term represents the number of heterozygous loci and the exponent of the second term represents the homozygous loci. It can be helpful to reconstruct the formula in the form $(a + b)^n$ where $a = 1$ and $b = 2^m - 1$. Consider the example of three loci in the F_4 generation. The formula expansion would be $(a + b)^3 = a^3 + 3a^2b + 3\,ab^2 + b^3$. The first term ($a^3$) represents the number of individuals heterozygous at all three loci, the second term ($3a^2b$) includes those that are heterozygous at two loci and homozygous at one, and so on. We must now calculate the values of a and b. Here a is always 1, and b in this case equals $(2^m - 1)$ or 7. Thus the formula can be computed as follows: $1^3 + 3(1)^2(7) + 3(1)(7)^2 + 7^3$ and is summarized as:

3 heterozygous, 0 homozygous :	1
2 heterozygous, 1 homozygous :	21
1 heterozygous, 2 homozygous :	147
0 heterozygous, 3 homozygous :	343
	512

The high degree of homozygosity resulting from self-fertilization is again pointed out by this example since only 1/512 or 0.2 percent are still heterozygous at all three loci in the F_4.

Cross-Fertilization

Cross-fertilized population dynamics are expressed in the Hardy-Weinberg law that states that genotype and phenotype frequencies are maintained in equilibrium after one round of random mating, if several assumptions are satisfied. These include completely random mating, no differential selection, no unidirectional mutation rates, no immigration or emmigration of alleles, large numbers of individuals in the population, and diploid composition. While all assumptions are not fulfilled in every population, Hardy-Weinberg equilibrium principles can generally be applied to many cross-fertilized species.

To illustrate, we will again start with a group of individuals heterozygous *(Aa)* at one locus. The intermating of these individuals will produce the familiar 1*AA* : 2*Aa* : 1*aa* ratio in the next generation. If the population is completely cross-fertilized, then each gamete has equal potential of combining with any other gamete in that round of mating. In our population the *A* gamete will be produced with exactly the same frequency as *a*. When these are mated at random a 1*AA* : 2*Aa* : 1*aa* ratio is again produced. This pattern will be repeated for unlimited generations and for any number of loci. Thus, in a cross-fertilized species a large number of heterozygotes are maintained and recessive alleles can remain hidden in the heterozygous genotypes.

ASEXUAL REPRODUCTION

Many plants with high commercial value are reproduced by asexual means—that is, without the combination of gametes. Fruit trees are traditionally multiplied by cuttings. Strawberries are reproduced by runners. The eye of the potato is planted to produce the new crop. Perennial grasses and legumes are often cloned vegetatively by division of the crown. Roses are sold primarily as elite shoots grafted to hardy rootstocks. Indeed, many valuable crops such as seedless grapes and oranges would be lost without the benefit of vegetative reproduction. Considerable research has been conducted on the economic aspects of asexual reproduction in plants. Wright (17) describes plant reproduction techniques with special emphasis on asexual propagation.

The above examples represent those cases where obvious vegetative tissue is used to produce the next generation. Apomixis is another form of asexual reproduction in which a seed is produced without the combination of male and female gametes. Several varitions of apomixis occur, but the essential component of all types is that of a female gamete, either reduced or unreduced, being formed and developing into a seed. A new individual is formed without genetic contribution from the male. Apomixis is often confused with sexual reproduction because seed is formed in both cases and extensive research may be required to determine if the seed is produced sexually or apomictically. Genetic markers are frequently used to identify the amount of apomixis present. Apomixis is particularly attractive in vegetative plant propagation since the seed can be stored and transported much more easily than other vegetative propagules. Apomixis is genetically controlled, and a summary of the subject is provided by Khaklov (2).

Another form of vegetative seed production is nucellar embryony, common to many of the Citrus species such as oranges. Here somatic cells in the nucellus develop into seeds, but are entirely of the female genotype since they did not originate by fertilization. In some varieties very few sexual seedlings are ever recovered. The stimulation of fertilization may be required to initiate nucellar embryony, but following fertilization the nucellar cells become aggressive and compete successfully for space with the zygotic embryo. Both environment and genotype appear to participate in the control of this character.

These are only a few of the many variations of vegetative reproduction, yet they serve as examples with interesting ramifications in plant breeding. The genetic implications of vegetative reproduction are clear. If there are neither mutational events nor the union of male and female gametes, genotypes will remain constant. This offers some distinct breeding advantages since any desirable genetic combination can be fixed immediately and propagated indefinitely. The highly desirable Russett Burbank potato, for example, was developed prior to 1890 but still occupied 40 to 45 percent of the total U. S. acreage in 1978 with apparently very little genetic change. On the other hand, breeders of vegetative crops are sometimes faced with very difficult problems. Since zygote formation through sexual union is not as rigidly enforced in vegetative crops as in seed crops, genetic recombination may be difficult or impossible to accomplish. If recombination can be effected, however, genetic principles apply in exactly the same manner to sexually and asexually reproduced crops.

SUMMARY AND COMMENTS

The mode of propagation presents a unique set of challenges in each plant species. An understanding of the reproductive mechanism is important in producing new genetic combinations and taking advantage of genetic variability. Breeders, by necessity, have become very innovative in their crossing and plant reproduction techniques. Both sexual and asexual reproduction systems have their genetic advantages and limitations.

As a breeder of self-pollinated grass—wheat—I have not worked with many of the interesting flowering and pollination mechanisms available in other species. Students assure me that it is much more difficult to make a cross in beans than it is in wheat. Regardless of the difficulties, a sense of creativity exists for any breeder watching the formation of a new embryo from the deliberate mating of two parents with the potential to produce a new and better variety. This creativity is an important aspect of the plant breeding profession.

REFERENCES

1. *Advances in fruit breeding.* 1975. J. Janick and J. N. Moore (eds.), Purdue Univ. Press, West Lafayette, Ind.

2. *Apomixis and breeding.* 1976. S. S. Khaklov (ed.). Amerind, New Delhi.

3. Barrett, H. C., and T. Arisumi. 1952. Methods of pollen collection, emasculation and pollination in fruit breeding. *Proc. Am. Soc. Hort. Sci.* 59:259–262.

4. Erickson, E. H. 1975. Variability of floral characteristics influences honey bee visitations to soybean blossoms. *Crop Sci.* 15:767–771.

5. Falgri, K., and L. van der Pijl. 1971. *The principles of pollination ecology.* Pergamon Press, New York.

6. Frankel, R., and E. Galun. 1977. *Pollination mechanisms, reproduction and plant breeding.* Springer-Verlag, New York.

7. *Hybridization of crop plants.* 1980. W. R. Fehr and H. H. Hadley (eds.). Am. Soc. Agron., Madison, Wisc.

8. Mangelsdorf, P. C., and G. S. Fraps. 1931. A direct quantitative relationship between vitamin A in corn and the number of genes for yellow pigmentation. *Science* 73:241–242.

9. Pandey, K. K. 1960. Evolution of gametophytic and sporophytic systems of self-incompatibility in angiosperms. *Evolution* 14:98–115.

10. ———. 1968. Compatibility relationships in flowering plants. Role of the *S* gene complex. *Amer. Natur.* 102:475–489.

11. ———. 1977. Origin of complementary incompatibility systems in flowering plants. *Theor. and Appl. Genet.* 49:101–109

12. Percival, M. S. 1965. *Floral biology.* Pergamon Press, New York.

13. Poehlman, J. M. 1979. *Breeding field crops.* 2nd Ed. AVI, Westport, Conn.

14. Rhoades, M. M. 1936. The effect of varying gene dosage on aleurone colour in maize. *Jour. of Genet.* 23:347–354.

15. Uphof, J. C. T. 1938. Cleistogamic flowers. *Bot. Rev.* 4:21–50.

16. Westergaard, M. 1958. The mechanism of sex determination in dioecious flowering plants. *Adv. in Genet.* 9:217–281

17. Wright, R. C. M. 1975. *The complete handbook of plant propagation. Macmillian, New York.*

QUESTIONS

1. In a gametophytic incompatability system, what are the progeny from the following crosses?
a. ♀ S_1S_2 × ♂ S_3S_4 **Answer:** S_1S_3, S_1S_4, S_2S_3, S_2S_4
b. ♀ S_1S_2 × ♂ S_1S_3 **Answer:** S_1S_3, S_2S_3
c. ♀ S_1S_3 × ♂ S_1S_2 **Answer:** S_1S_2, S_2S_3

2. In a sporophytic incompatability system, what are the progeny from the following crosses?
a. ♀ S_1S_2 × ♂ S_3S_4 **Answer:** S_1S_3, S_1S_4, S_2S_3, S_2S_4
b. ♀ S_1S_2 × ♂ S_1S_3 **Answer:** No progeny
c. ♀ S_1S_3 × ♂ S_3S_4 **Answer:** No progeny
d. ♀ S_1S_3 × ♂ S_2S_3 **Answer:** S_1S_2, S_1S_3, S_2S_3, S_3S_3

3. Suppose a dominant allele for red (R) is linked *very closely* with the gametophytic incompatability allele S_1 and the white allele r is linked with S_2, S_3, and S_4.
a. What color would the S_1S_2 genotype be? **Answer:** Red.
b. What would the color segregation be in a cross $S_1S_2 \times S_1S_3$. **Answer:** 1 red : 1 white.

4. In a self-fertilizing population, what is the proportion of heterozygosity at one locus after seven generations of selfing including the F_1? **Answer:** 1/32

5. In a self-fertilizing population, what is the proportion of completely homozygous individuals in the F_6 if four loci are segregating between the parents?

Answer: $\frac{923{,}521}{1{,}048{,}576}$ or 88.07%

6. In a self-fertilizing population with five loci segregating, what proportion of the population in the F_5 will be heterozygous at three loci and homozygous at two loci?

Answer: $\frac{2{,}250}{1{,}048{,}576}$ or 0.21%

7

NATURAL GENETIC VARIATION

Without genetic variation, any adverse changes in the environment would doom a species to extinction in its natural habitat. This chapter deals with the occurrence of natural genetic diversity, the role it plays in evolution, and various systems for its collection, preservation, distribution and utilization.

ALLELIC VARIATION

Occurrence

Genetic diversity in the form of allelic variation is created by mutations. They occur spontaneously with frequencies that can vary depending on the locus itself and on genetic information in neighboring regions of the chromosome. Natural mutation rates for a number of different loci in corn ranged from 0 to about 500 mutations per 1,000,000 gametes according to Stadler (17), emphasizing the point that some loci are easily mutated while others are very stable. Mutation rates themselves are also under genetic control. McClintock (14) described genes in corn, named activators and disassociators, that controlled the mutation rates of other loci. The level of stability against mutation depended on the allelic condition of the control loci. Environmental factors such as radiation load and temperature shock commonly affect mutation rates. Energy sources are undoubtedly a major cause of many mutational events.

Whatever the cause, the net result is a changed DNA message, altered enzymes, and potential variation in the physiological mechanism to be evaluated in the natural selection process.

Value

The selective value of each mutation depends on the testing environment and the total genotype of the tested individuals. For example, a mutation resulting in a very high level of cold tolerance may have no value or even be a disadvantage in a warm climate. Yet this same mutation may have high selective advantage in a cold environment. In combination with short photoperiod requirements dictated by alleles at other loci it may have little value, but may be very valuable in a genotype requiring a long photoperiod. Often, the selective advantage or disadvantage of a mutation is not particularly dramatic, but instead may represent the potential for slight deviations from the average performance of the population.

A high proportion of successful mutations appear to be recessive in nature. This may be due to the long evolutionary history of plant species where well-adapted genotypes have survived the selection process. Assuming that cross-fertilization is the natural reproductive system, dominant mutants would appear immediately in the phenotypes and be selected or eliminated rather quickly. Recessive mutations, on the other hand, could survive and propagate in a population in the heterozygous condition without producing significant phenotypic effects. They would be available over a long time period for potential recombination in many genotypes with subsequent selection advantage evaluation.

The value of a mutation may vary over time because of changes in the environment. A disease resistance gene would have little or no selective advantage unless the disease becomes a problem. Then the resistance gene becomes very valuable.

FUNCTIONS OF VARIATION

Natural Evolution

Darwin (4) recognized the value of variation in the evolutionary process. While he was not clear on the hereditary mechanisms, his classic works stressed that natural evolution proceeds because heritable variations exist. DeCandolle (5) developed an extensive document in 1886 that describes the variations associated with the evolution of cultivated crop species. Notice that both works were produced prior to the rediscovery in 1901 of Mendel's reports on inheritance principles.

Mutations, in themselves, are important and are continually being tested for selective value. Equally important is the continual generation of new genotypes, unique combinations of alleles produced through recombination. The following discussion deals with systems for producing new genotypes.

Recombinations Within a Species Genetic recombination depends on a high degree of crossing between genetically different individuals. The highest levels of outcrossing occur among plants that contain protection mechanisms against self-fertilization. Stebbins (18) suggests that self-fertility is an evolutionary refinement from populations that were originally outcrossed. From the discussion on the genetic implications of selfing and crossing, a more complete mixing of genes and alleles can be visualized in the cross-fertilization system. Stebbins points out that a number of species have evolved quite satisfactorily on a selfed basis, but even here a low level of outcrossing occurs to allow genetic recombination.

A mutation may occur as a single unique event, or it may take place several times, especially in a large population. Following the mutation, DNA duplication allows many copies of this new allele to be produced and transmitted through the reproductive cells. As plants outcross, a number of progeny are produced that may be carrying the mutant allele in the heterozygous combination with other alleles. As plants continue to reproduce and intermate, the mutation undergoes recombination with different alleles at many loci. This allows nature to apply selection pressure on various genetic combinations over long periods of time and identify those genotypes that can best survive. The mutation system is undoubtedly the most common and powerful source of genetic variation available for evolution.

Recombination Between Species While the blending of the genetic variability within a species is important, the incorporation of variability from different species and genera also represents a valuable tool in evolution. Generally speaking, the wider the phylogenetic distance between the parents in a cross, the lower the success rate in generating viable and fertile progeny. Fertility is critical because effective genetic recombination has not been achieved unless the genes can be passed from generation to generation.

Problems in wide crosses commonly arise from three causes. First, fertilization may be difficult to accomplish because the pollen tube is not initiated or is unable to complete the journey to the egg. Second, the developing zygote may abort very quickly. Third, the hybrid may form and develop successfully but be sterile because of difficulties in producing viable gametes. Sterility in the hybrid comes about primarily because the nonhomologous chromosomes in the different species will not pair, resulting in abnormal chromosome distribution to the gametes.

All three problems represent barriers to interspecific gene transfer, but nature does allow some genetic interchange by two methods. The first of these is called introgression. Consider the situation where an F_1 has been produced between two individuals from different species cohabiting the same area at the edge of their respective habitats. The F_1 is viable but produces only a very few functional gametes with chromosomes from both species. If the F_1 mates with one of the parent species and produces a viable offspring, the chances of functional gametes increase. This results from the increasing proportion of chromosomes from one parent with associated improvement in pairing and movement. As repeated crosses are made between each generation of hybrids and the recurring parent, the chromosomes of the nonrecurring parent are gradually eliminated. The possibility exists, however, for a small amount of genetic information to be retained from the nonrecurring parent. This is done either by retention of one or a few whole chromosomes, or by the incorporation of a part of one chromosome into the recurring parent chromosomes by breakage and rehealing. A whole chromosome can be added to the total complement of the recurring parent (an addition) or it can take the place of one of the recurring chromosomes (a substitution). When a chromosome breaks, an additional piece from some other source can be inserted or added before the break heals (a translocation). The crossing process may be repeated several times until the genetic contribution from one parent is very small. The net result, however, is a genetic interchange between species. Anderson (1) describes the mechanics and details of introgression.

Apparently introgression has been a valuable evolutionary tool occurring fairly commonly in nature. The progeny of crosses between species can often be found in overlapping ecological regions of different species. Zohary (23) cites evidence that introgression has operated extensively in the evolution of sunflowers, maize, wheat, rye, tomatoes, and several other crops.

Polyploidy is another system for combining variability between species. Principles of the mechanisms have been discussed in Chapter 5. Polyploidy has the advantage of much more nearly normal meiosis—providing amphidiploids are formed. Allopolyploidy has played a significant role in the evolution of many cultivated species. Many different ploidy levels have been generated naturally but Swaminathan (19) reports that the octaploid is often suggested as the ceiling for flowering plants. Diploidization, the stabilization of diploid-like chromosome synapsis and movement in meiosis following polyploid formation, is a common evolutionary occurrence. This is undoubtedly because of sterility problems associated with abnormal chromosome configuration in meiosis I.

Autopolyploidy can also be an important factor in plant evolution. According to Stebbins (18), it seems to exist in cross-fertilized species but not in those that are selfed. Again, diploidization is required for normal gamete formation. Genetic control of pairing plays a key role in diploid-like chromosome performance of an autopolyploid. Examples of autopolyploid species include the potato, alfalfa, and western wheatgrass.

Human Directed Evolution and Plant Migration

Human intervention has contributed significantly to crop evolution and is well described by Harlan (12). As humans began to establish cultures and domesticate plants, they selected those genotypes that best satisfied their needs. Harlan stresses that high levels of productivity were not necessary but stability of production was desired. Instability during brief periods of abnormal environmental fluctuations such as severe drought or cold generally resulted in disaster.

Such characters as taste, color, lack of shattering, large seed, and ability to establish a stand were important in the early domestication programs. Apparently, primitive cultures did not require the high levels of uniformity that we see in crops today. Many varieties, called land races, originate through individual farmer or small geographic area propagation and are quite diverse in their genetic composition. In fact, land races serve as good reservoirs of genetic variability for germplasm collection programs.

As people migrated from one area to another, plants or seeds were carried along and tested in the new environment. This process, called introduction, is really a form of plant breeding since it offers the potential of using different genotypes to improve plant performance in a given environment. Many introduced species and varieties were poorly adapted and failed to survive. There were some noticeable successes, however. Hard red winter wheat, which dominates wheat production in the Great Plains, was introduced by Mennonites immigrating from Russia. The potato, which has represented a major food supply source for much of Europe and especially the British Isles, was introduced from North America after originating in South America. The origin of many other commercial crop species will include migration and introduction as part of the recent evolutionary process.

Use of Variability as Breeding Material

Natural occurring genetic variability is the grist for the mill of any plant breeding program. The variation can be exploited, as early humans did, through simple introduction and selection techniques or it can be utilized in sophisticated crossing programs to produce new genetic combinations. The value of natural genetic variation to plant improvement is well documented in pest resistance (20) and many other characters.

Much concern has been expressed by the scientific community in recent years regarding the concept of genetic vulnerability. This is the situation where, because of intensive farming practices and the widespread use of a very few highly productive genotypes, crop production is increasingly vulnerable to a catastrophe such as a race change in a disease organism, a shift in the weather pattern, or lack of fertilizer because of depleted energy supplies. Demonstrations of vulnerability are all too common. To be convinced, one has only to study the periodic stem rust devastations of the U. S. Great Plains wheat crops, the loss of 15 percent of the U. S. corn crop to Southern Corn leaf blight in 1970, the disease and insect problems that have regularly threatened the grape industry, or the disease-caused potato famine in Ireland in the 1840s.

Genetic vulnerability was covered extensively in scientific conferences held by the National Academy of Science-National Research Council in 1972 (9) and 1978 (2) and by the New York Academy of Science in 1976 (20). All conference participants emphasized the urgent need to find and preserve naturally occurring genetic variation in order to adequately respond to the problem of genetic vulnerability. This view is supported by Creech and Reitz (3) and Harlan (11). The following sections discuss germplasm collection, preservation, and distribution. The mechanics of utilizing this variation are examined in chapters on specific breeding programs.

Collection

Before considering germplasm collection, we must ask the question: "Do genes and genotypes actually become lost or extinct?" The evidence strongly suggests an affirmative answer (7). The problem arises from the rapid changes that are taking place in the environments of naturally occurring populations. Many of these changes are associated with human activity such as cultivation, road and urban building, introduction of chemicals, and the recent switch from land races to modern high yielding genotypes. Reitz (16) reports that Russian scientists recently were unable to find a certain wheat species in geographic areas where it was obtained in the 1930s. The wild diploid potato species are disappearing from Bolivia, Argentina, and the Venezuelan Andes, according to Ochoa (15). These are two examples of genes and genotypes being lost each year, indicating that some type of preservation is required.

Human interest has always been stimulated by variation. Collections of all kinds have become an international phobia and plant materials are no exception. People with widely diverse interests have historically collected complete plants or their reproductive mechanisms such as seeds or vegetative parts. Much of the early work, however, dealt with taxonomic classification, rather than collections for the preservation of genes. One of the foremost pioneers in the science of germplasm collection was N. I. Vavilov, a Russian geneticist and agronomist for whom the enormous N. I. Vavilov All-Union Institute of Plant Industry in Lennigrad was named.

Vavilov, after extensive plant collection trips in the early part of the twentieth century, proposed the concept that cultivated crop species had centers of origin and that these centers were identified by the greatest amount of genetic diversity. This concept has been very useful in exploring for genetic diversity. Harlan (10) points out, however, that there are centers of origin and centers of diversity and the two may not coincide. In fact, he proposes that in some species the activities of plant domestication have gone over such a large geographic area that the term "center" does not apply but instead these should be called noncenters. While the evolutionary picture may always remain cloudy, it is evident that genetic diversity is available for collection and preservation.

To this end, many governments around the world have sponsored plant expeditions for the purpose of germplasm collection. In the U. S., for example, 65 foreign explorations were undertaken by the USDA during the period 1946 to 1971. Scientists in these expeditions travelled to many parts of the world in search of genetic variability in a wide array of plant species (8).

Preservation

Having located the genes, the next major considerations include identification, cataloging, and preservation.

Within the U. S. an extensive system of germplasm management has been established (21,22) and is summarized in Figure 7.1. The USDA Germplasm Resources Laboratory located in Beltsville, Md., serves as the national focal point for introduction, documentation, initial distribution, and exchange of germplasm. If live vegetative plant material is a requirement in propagating the genes, an immediate problem is presented in transportation and

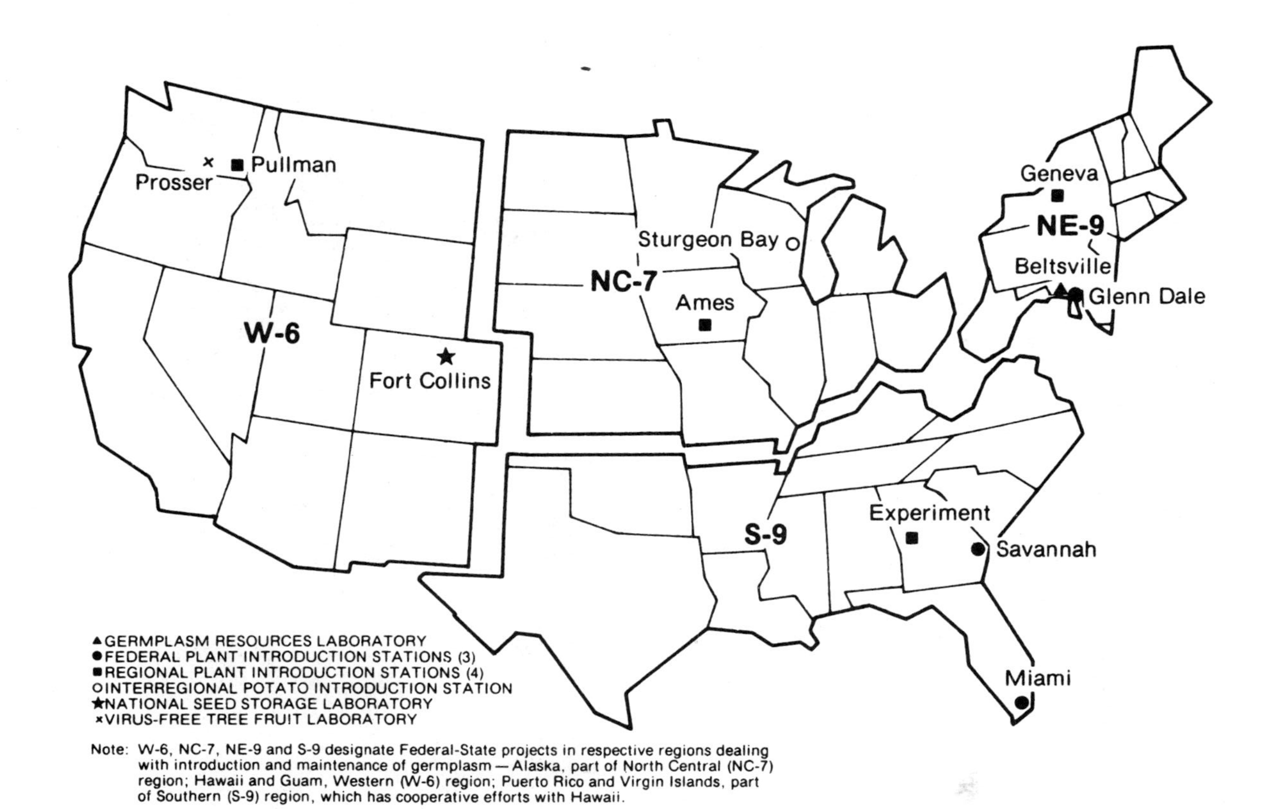

Note: W-6, NC-7, NE-9 and S-9 designate Federal-State projects in respective regions dealing with introduction and maintenance of germplasm — Alaska, part of North Central (NC-7) region; Hawaii and Guam, Western (W-6) region; Puerto Rico and Virgin Islands, part of Southern (S-9) region, which has cooperative efforts with Hawaii.

Figure 7.1. Major stations of the U.S. National Plant Germplasm System. (Courtesy of USDA-SEA.)

possible transmission of insects and diseases. To respond to this problem in the U.S., a quarantine system has been established so that all material is propagated under strictly isolated conditions, for some period of time, varying with the species, to insure that no undesirable diseases or insects are present. The Beltsville laboratory is the lead organization in the operation of the quarantine and detention program. It also maintains responsibility for the world collection of over 85,000 stocks of small grains. Included in this group are wheat—35,000, barley—22,500, rye and triticale—1,600, rice—10,000 and oats—17,000. Figure 7.2 shows some of the world collection of barley being grown for classification and seed replenishment. This must be done periodically so adequate seedstocks will be available to satisfy breeder and geneticist requests.

In addition, USDA-State Agricultural Experiment Station Cooperative Regional germplasm evaluation and maintenance stations are located at Ames, Iowa; Experiment, Georgia; Pullman, Washington; and Geneva, New York. Each station has been designated to receive, multiply, maintain, and distribute introduc-

Figure 7.2. A portion of the barley world collection. Each row is a different genotype. Note the diversity of phenotypes with respect to height, maturity, and head characteristics.

tions of certain species. Interregional operations include the Potato Introduction Station at Sturgeon Bay, Wisconsin, and the Virus-free Deciduous Tree Fruit laboratory at Prosser, Washington. Completing the introduction stations are federally operated laboratories at Glenn Dale, Maryland; Savannah, Georgia; and Miami, Florida.

Approximately 20 plant material centers around the U. S., each with geographic areas of responsibility, are operated by the Soil Conservation Service. The centers have the function of evaluating and propagating superior germplasm collected by field scientists for purposes of conservation, aesthetic value such as roadside and park revegetation, and reclaiming disturbed lands. These centers have served as a valuable source of germplasm for breeders in related programs.

The National Seed Storage Laboratory in Fort Collins, Colorado became operational in 1958. The purpose of this facility is to store all collected germplasm in a viable condition under controlled temperature and humidity. Over 110,000 introductions are currently housed there. Problems of storage include decrease in viability over time, genetic change while in storage, and reproduction of material to maintain viability. Although some germplasm can be stored in the seed condition for several years, others may require replenishing annually. Research is being conducted on systems of storage such as liquid nitrogen to prolong longevity.

A large number of repositories exist around the world, each with its own groups of crops for use in that particular geographic region. Some are governmentally sponsored, others are operated with international funding, and a few are supported by seedspeople and private companies. Taken together, a large network exists to collect and maintain genes. Frankel (6) gives a comprehensive listing of storage centers. To consolidate and emphasize worldwide efforts in the conservation of plant genetic resources, the International Board of Plant Genetic Resources was founded in 1974. IBPGR, located in Rome, is supported by a group of organizations with international agricultural development interests.

Normally, we think of storing germplasm by conventional methods in physical facilities of some kind. Populations should also be maintained in their natural habitat to allow for normal mutation and natural selection. This generally calls for the establishment of federal refuges or some type of isolation from the human element. A limited number of areas have been so designated but this a difficult program to implement in many primitive regions where the greatest germplasm diversity exists.

Distribution and Information Systems

To achieve maximum value, genes must be available for use by a large number of scientists. Generally, within the U. S., germplasm is available to anyone, public or private, who submits a request to the respective distribution centers. Approximately 190,000 samples are distributed by the U. S. National Plant Germplasm System (21) each year. Normally, requests come from breeders looking for a particular character expression. They may request, for instance, a portion of a world collection to screen for some form of disease resistance. The information obtained in the screening process is then returned to the collection curator and entered into the record for future use.

A serious problem occurs when the information has not been described in a uniform manner. What may be clear to one evaluator may be quite confusing to another. With the sophistication of computers, information systems have been initiated that will assist in description techniques and allow breeders to request and utilize germplasm with specific groups of characters. Hersh and Rogers (13) describe a system that is now being developed. The strength of any information system lies in accurate uniform classification of the germplasm and the practicality of use for the breeder.

On an international basis, germplasm exchange is encouraged and utilized by breeders. Recent relaxation of political barriers has lead to a much greater level of exchange than before and it is expected that even more will take place in the future.

A final method of gene distribution involves international plant improvement programs. Examples include the International Maize and Wheat Improvement Center (CIMMYT) in Mexico, the International Crops Research Institute for Semiarid Tropics (ICRISAT) in India, the International Rice Research Institute (IRRI) in the Philippines, and the International Center for Tropical Agriculture (CIAT) in Columbia. A common practice of these organizations is to assemble large blocks of exotic germplasm and incorporate them into adapted varieties through massive crossing programs. Following hybridization, gene pools are distributed as segregating populations to many individual programs throughout the world. This represents an efficient system that removes much of the financial burden of germplasm collection, evaluation, and incorporation from localized programs with limited resources.

SUMMARY AND COMMENTS

The need to preserve existing genes for use, either currently or at some future time, is critical. An extensive germplasm system has been generated in the U. S. and around the world. The value of genes is not always apparent and the cost of collecting, classifying, storing, maintaining, and distributing is sometimes difficult to justify in the eyes of the taxpayer.

Two personal experiences, both involving plant diseases, can serve as testimony to the value of germplasm. When I was growing up on a wheat farm in North Dakota, our crops were periodically devastated by stem rust to such an extent that the economics of our operation was highly uncertain. Breeding programs for rust resistance have provided protection so that serious outbreaks of the disease have not occurred in the past 20 years. The second instance took place while I was working as the winter wheat breeder at Montana State University. My predecessor, Dr. E. R. Hehn, and I were confronted with two serious diseases, stripe rust and dwarf smut, for which there seemed to be little genetic or cultural control. In the early 1960s we experienced a severe outbreak of both diseases and the world collection was screened for resistance. Contained in the collection was a genotype collected from Turkey by Dr. Jack Harlan in 1948 and identified as P.I.178383. It was a miserable plant type, since it was tall and susceptible to lodging, without winter hardiness but very difficult to vernalize, and with very undesirable milling and baking properties. It was, however, identified as having excellent resistance to stripe rust and dwarf smut as well as several other diseases. The undesirable genotype turned out to be one of the most valuable lines in the collection for us, and was used in developing resistant varieties in our program and in many others throughout the Pacific Northwest. It was so heavily used, in fact, that genetic vulnerability again became a possibility.

Breeders are continually looking for better genes. As in most good detective stories, the final identification of the wanted individual, while very exciting and rewarding, comes from considerable routine effort.

REFERENCES

1. Anderson, E. 1949. *Introgressive hybridization.* Wiley, New York.

2. *Conservation of germplasm resources, an imperative.* 1978. Natl. Res. Coun., Natl. Acad. Sci., Washington.

3. Creech, R. G., and L. P. Reitz. 1971. Plant germplasm—now and for tomorrow. *Adv. Agron.* 23:1-49.

4. Darwin, C. R. 1859. *On the origin of species by means of natural selection.* J. Murray, London.

5. DeCandolle, A. 1886. (3rd printing 1967). *Origin of cultivated plants.* Hafner, New York.

6. Frankel, O. H. 1973. *Survey of crop genetic resources in the centers of diversity.* First report AGP-CGR 73/7. FAO-IBP report. FAO, Rome.

7. *Genetic resources for today and tomorrow.* 1975. IBP Programme 2, O. H. Frankel and J. G. Hawkes (eds.). Cambridge Univ. Press, Cambridge.

8. *Genetic resources in plants—Their exploration and conservation.* 1970. International Biological Programme Handbook No. 11, O. H. Frankel and E. Bennett (eds.). Blackwell Scientific, Oxford.

9. Genetic vulnerability of major crops. 1972. Natl. Res. Coun., Natl. Acad. Sci., Washington.

10. Harlan, J. R. 1971. Agricultural origins: centers and noncenters. *Science* 174:468-474.

11. ———. 1975. Our vanishing genetic resources. *Science* 188:618-621.

12. ———. 1975. *Crops and man.* Am. Soc. Agron., Crop Sci. Soc. Am., Madison, Wisc.

13. Hersh, G. N., and D. J. Rogers. 1975. Documentation and information requirements for genetic resources applications. pp. 407-446. In O. H. Frankel and J. G. Hawks (eds.). *Crop genetic resources for today and*

tomorrow. International Biology Programme No. 2. Cambridge Univ. Press, Cambridge.

14. McClintock, B. 1950. The origin and behavior of mutable loci in maize. *Proc. Natl. Acad. Sci.* 36:344-355.

15. Ochoa, C. 1975. Potato collecting expeditions in Chile, Bolivia, and Peru, and the genetic erosion of indigenous cultivars. pp. 167-173. In O. H. Frankel and J. G. Hawkes (eds.). *Crop genetic resources for today and tomorrow.* International Biological Programme 2. Cambridge Univ. Press, Cambridge.

16. Reitz. L. P. 1976. Improving germplasm resources. pp. 85-97. In F. L. Patterson (ed.). *Agronomic Research for Food.* Am. Soc. Agron. Special Pub. 26.

17. Stadler, L. J. 1942. "Some observations on gene variability and spontaneous mutation," The Spragg Memorial Lectures on Plant Breeding (Third Series), Michigan State College.

18. Stebbins, G. L. 1957. Self fertilization and population variability in higher plants. *Am. Nat.* 91:337-354.

19. Swaminathan, M. S. 1970. The significance of polyploidy in the origin of species and species groups. pp. 87-96. In O. H. Frankel and E. Bennett (eds.). *Genetic resources in plants—their exploration and conservation.* International Biological Programme Handbook No. 11. Blackwell Scientific, Oxford.

20. The genetic basis of epidemics in agriculture. 1977. *Annals of New York Acad. Sci.,* Vol. 287.

21. *The National Plant Germplasm System.* 1977. USDA Program Aid No. 1188.

22. *The National Program for conservation of crop germplasm.* 1971. Sam Burgess (ed.). Univ. of Georgia Printing Dept., Athens.

23. Zohary, D. 1970. Centers of diversity and centers of origin. pp. 33-42. In O. H. Frankel and E. Bennett (eds.). *Genetic resources in plants—their exploration and conservation.* International Biological Program Handbook No. 11. Blackwell Scientific, Oxford.

8

VARIABILITY IN BIOLOGICAL SYSTEMS

Differences will always exist among individuals in a population. This is true whether it is a group of students in an exam or plants in a field. In a breeding program, variation and its identification are the key to success and the cause of frustration and failure.

Theoretically, all the variation in biological systems can be divided into two categories: heredity (genetics) and environment. This chapter considers these two sources of variation and their measurement and interpretation.

VARIATION DUE TO ENVIRONMENT

Environment is the sum total of all things to which an organism is exposed. To illustrate, suppose we are standing on the edge of an apple orchard just prior to fruit harvest. Consider the environmental conditions, past and present, that are reflected in the yield of a single tree along the road on the edge of the orchard. Some obvious items include radiation duration and intensity, frequency and amount of moisture supplied (both rainfall and irrigation), and the nutrients provided either through fertilization or from the soil. Add to this list the amount and direction of field slope, presence of diseases and insects, and availability of bees at pollination time. Less obvious factors include the passage of cars and trucks along the road and the frequency and pattern with which tillage equipment has passed by the tree. These and many other environmental effects add together and interact among themselves and with the genotype of the tree to produce final yield.

Now consider a tree six rows away from the road. Assume that both were produced by cuttings from one original parent and, therefore, have the same genotype. The general list of environmental factors reads very much the same as for the first tree, yet the yield is quite different. Reasons for the difference are numerous. For example, the second tree is surrounded by trees on all four sides while the first has neighbors only on three sides. It is not as close to traffic and is thus exposed to a slightly different atmospheric composition, it receives a different frequency and pattern of tillage equipment traffic, and it was planted in the edge of an area where an old corral had once been.

Since the genotypes in this example are identical, any differences in yield are credited entirely to environment. We touched on only a few of the large magnitude of easily identified environmental influences. Others are so subtle that it is almost impossible to identify and measure their impact. In fact, a completely uniform environment can never be produced even under the closely controlled conditions of growth chambers and greenhouses that may be relatively free of environmental fluctuations. Environmental variations are always present and must be dealt with in any scientific endeavor including a plant breeding program.

Each breeder is faced with an array of environments in which his or her breeding program is to achieve results. Under some production conditions the environment is relatively uniform and highly controlled. The best example of this is a desert situation in which plant production is based almost entirely on irrigation as the main water supply. Radiation is relatively constant because of a small amount of cloud cover, temperatures fluctuate on a highly predictable basis, and the environmental variations from day to day are quite minimal. Many variables can be controlled with a high degree of accuracy. On the other hand, dryland production systems represent a highly unstable environment. Precipitation events with variable intensity and duration, large and not entirely predictable fluctuations in temperature, high hazard events such as hail storms, and intermittent wind with rapid intensity changes add up to an environment that is highly variable. This situation is often characterized by averages that do not reflect the range and fluctuation of individual values. The breeder must be aware of the environmental characteristics in which the crop is to be improved genetically. Adequate understanding of the environmental composition will be reflected in decisions on genotype measurement mechanics and interpretation of results.

VARIATION DUE TO GENOTYPES

If the character expression of two individuals could be measured in an environment exactly identical for both, differences in expression would result from genetic control. Genetic variability is of prime interest to the plant breeder because proper management of this diversity can produce permanent gain in the performance of the plant.

Genotypic variation can be divided into the two broad general categories of qualitative and quantitative inheritance, with considerable overlap between the two.

Qualitative Inheritance

Qualitatively inherited traits are those in which the phenotypes can be distinctly separated into discreet categories or classes that do not overlap. With these traits the classification of individuals is very easy and there is little variation from environmental interaction with the genotype. The characters Mendel investigated were of this type. There was little doubt, for instance, about the classification of seed with respect to wrinkled or smooth surface and the phenotype would have had approximately the same expression over a wide range of environments.

To contrast qualitative and quantitative inheritance, a theoretical system will be used in which height is the character under consideration. One homozygous parent will have a value of 12 units and the other a value of 24 units. Various types of F_2 distributions will be generated, depending on the inheritance system imposed.

A qualitative situation is illustrated in Figure 8.1*a.* Here we have an F_2 that has been divided into two widely separated phenotypic classes. A single locus controls the character, the gene action is completely dominant, and there is little, if any, environmental interaction. The frequency distribution in Figure 8.1*b* shows a slight modification of the population distribution because of a

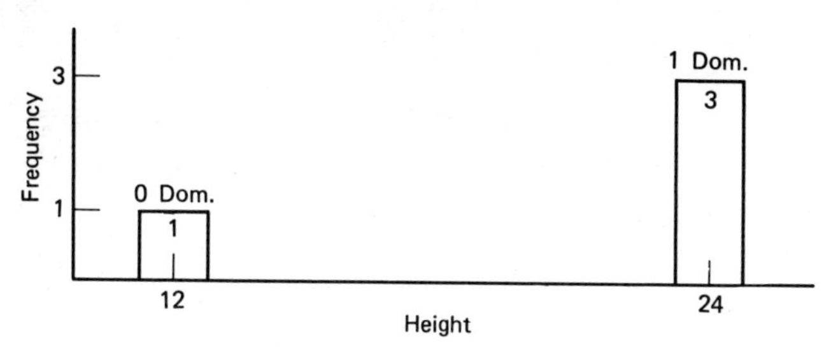

(a) One locus, dominant phenotype is 12 units taller than recessive.

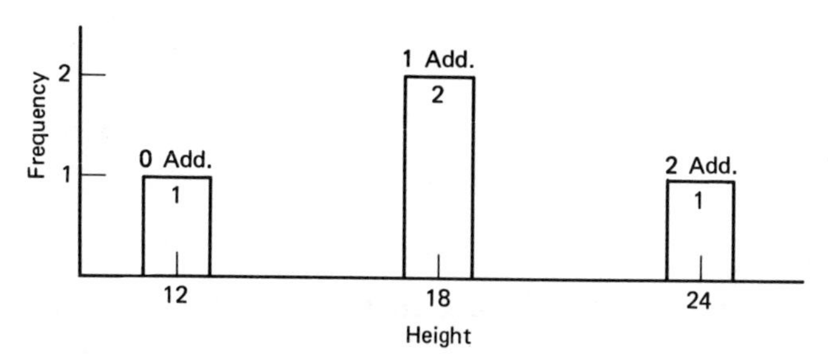

*(b)** One locus, completely additive with each additive allele adding 6 units.

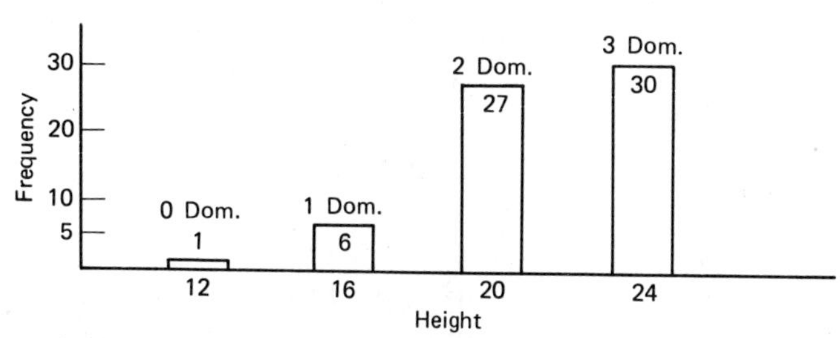

(c) Three loci, complete dominance within each locus and each dominant locus adding 4 units.

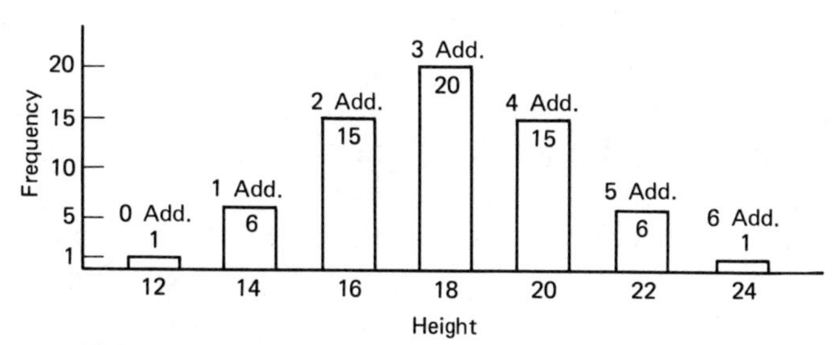

(d) Three loci, completely additive within and between loci with each additive allele adding 2 units.

Figure 8.1. Effects on F_2 population distributions with dominant and additive gene action.

change in gene action. There is still only one locus controlling the character but the gene action is now completely additive. There are three phenotypic categories in the F_2, each one quite distinct from the others. Note that we are still dealing with qualitative inheritance since the phenotypic categories are distinct and there is no difficulty in placing the progeny in their respective classes, even though more classes with less distance between them are expressed.

Quantitative Inheritance

Quantitatively inherited traits are generally characterized by the lack of discreet classes. Instead there is a great deal of overlapping from one class to another. The gradation occurs for one or both of two reasons. First, the genetic complexity governing the trait is increased by introducing interlocus interaction and additional numbers of loci. In Figure 8.1*c* the number of loci has been increased to three, with dominance between alleles within each locus and additivity among loci. Each locus that contains a dominant allele adds four units of height, with the heterozygous and homozygous condition at a locus having equal value. Four phenotypic categories occur and the frequency of each category depends on the number of loci where a dominant allele is present. The distance between each category has decreased to four units.

In the population distribution given in Figure 8.1*d,* three loci are still independently assorting, but the gene action has been changed to that of complete additivity within and between loci. Each additive allele adds two units of height. Notice that the increment of distance between each phenotypic class has been reduced to two and seven classes appear in the total distribution. The net effect is to bring the classes closer together as we add more loci and/or alter the gene action.

No environmental interaction has been added to the distribution patterns up to this point. All genotypes have been allotted distinct phenotypes and have been correctly classified. If, however, we introduce fluctuations in phenotypic expression due to environmental modifications, class separations become less and less, and are finally represented as a continuous set of values on the horizontal axis. Note that the population in Figure 8.1*d* is distributed normally because of genetic control. Normal distributions are also achieved if there is no genetic variation and the individuals measured represent a random sample from a large population. As inheritance systems become more complex and environment begins to play a larger role in character expression the interpretation of data becomes more difficult.

Qualitative and quantitative inheritance have been contrasted by widely differing population patterns. There is, of course, a large intermediate group that is difficult to categorize. This group represents the transition zone between the two inheritance systems and includes intermediate types of inheritance with environmental interaction that allows only some of the genotype expression in the phenotype. Many of our economically important plant traits fall into this group and the breeder must know how to interpret phenotypic values in terms of potential genetic gain.

Heritability

We have shown that the total variation in a population is the result of a combination of genotypic and environmental effects. The proportion of variation due to each source is of importance in plant breeding, since the amount of genetic variation is critical in achieving genetic gain. The proportion of total variation caused by the genotype is called heritability and can be described by the general formula $h = Vg/(Vg + Ve)$ where Vg is the genetic variation and Ve is the environmental variation. Heritability in the broad sense encompasses all types of gene action including dominance, additive, and epistasis. Heritability can theoretically range from one where all variation is genetic, to zero where all variation results from the environment. Actual heritabilities will fall somewhere between these two extreme values. Now return to Figures 8.1*a–d*. In these examples the character would have a heritability of one since the individuals were categorized correctly into their respective phenotypes, and the genotype was the exclusive cause of the phenotype. Gene actions of dominance and additivity only have been illustrated in these examples. Epistasis would also have specific effects on the frequency distributions.

As environmental effects are added into the total variation, the proportion of control resulting from genetic factors decreases and consequently heritability decreases. As this happens, the correct identification and measurement of each genotype becomes more difficult and genotype classification errors begin to appear. The difficulty in making breeding progress results from the lack of genetic variability, large environmental effects on phenotype, or some combination of both. It is very difficult to determine the presence, amount, or type of genetic variability if phenotypic expressions are strongly influenced by the environment.

Considerable research effort has gone into estimating the heritability of many economic traits. Heritability estimates call for the partitioning of the estimated variance into its genetic and environment components and this is often further divided into total genetic (wide sense) and additive (narrow sense) values. Several methods of heritability estimation are available and estimated heritability values for a single character will change depending on the statistical analysis technique used and the environment in which the experiment is conducted. Many estimates involve parent-offspring relationships. Allard (1) provides a summary of heritability concepts. With an exhaustive statistical analysis, a reasonably complete description of the gene action can often be accomplished so that additive, dominant, and epistatic effects are identified.

The most frustrating thing about heritability is that many of our commercially important traits, such as yield, certain types of disease resistance, and quality characteristics are often in the low heritability category. This means that there is a great deal of environmental influence on their expression, and genetic gain is reduced. On the other hand, as long as genetic variability exists, some progress can be expected.

MEASURING VARIABILITY

The measurement and evaluation of variability becomes extremely important in understanding the genetic control in biological systems. Several techniques are available and that choice depends on the goals, and the environmental circumstances of the breeding program. Every experiment has an underlying hypothesis and statistical concepts are used to estimate the probability of a proposed hypothesis being valid or invalid. Unfortunately, there is no way to be absolutely certain that the conclusion reached is correct.

The validity of statistical test results depends on employing the proper experimental design, and the use of accurate data collection methods. Erroneous conclusions from an improper test can be more serious than never having conducted the experiment, since each breeding program decision based on a false assumption represents a serious loss of resources in the total program. Decisions by breeders must be based on the most complete and accurate information available. However, since genetics and breeding are dynamic areas of science, the information picture will never be complete. The following sections consider several techniques of variation evaluation.

Genetic Ratios

Many plant breeding experiments evaluate the genetic control of characters including numbers of genes, types of gene action, and linkage by using ratio analysis with the chi-square test discussed in Chapter 3. Here the relationship between the actual data obtained and proposed ratios based on the hypothesis for some type of genetic control is studied. Conclusions are ultimately based on deviation magnitudes and patterns.

Ratio experiments are designed to minimize the environmental effect and maximize the genetic expression. They usually include some standard (check) nonsegregating genotype that is used to measure environmental variation. If the check genotype has a very consistent performance then the variation in the segregating individuals is likely the result of genotypic differences. The experiment may include several segregating generations. F_2's are commonly used, and in some cases F_3 families are also measured to verify the F_2 observations as Mendel did in describing his genetic mechanisms. Relating parent and progeny performance patterns is also used to identify and measure heritability. Each individual in the test represents an independent value, so the number of individuals must be large enough to provide a good estimate of the true genetic picture. The determination of the appropriate number may be a function of preliminary investigations.

As in all other statistical systems, we are dealing only in probabilities of hypothesis acceptance or rejection, and are not given absolute assurances that the conclusions are correct. There will always be some environmental effects compounding our data to obscure absolute answers.

Performance Comparisons

The breeder is commonly concerned with performance comparisons of genotypes. Will one variety yield more or have better quality properties or be more responsive to improved environmental conditions than another? These questions are often answered using the type of experimental design that will measure genotype characteristics and make comparisons on a statistical basis, while keeping the amount of environmental influence on the phenotypic values to a minimum. Any items placed in the test are called entries and may be different genotypes, fertilizer levels, or any experimentally controlled variable.

Almost without exception, two important statistical techniques are used in performance comparisons. Within each experiment each entry may be repeated or "replicated" several times to maximize the validity of the evaluation. Replications, often called blocks, allow for the reduction of statistical impact by some chance error associated with the physical location of the entry in the experiment. By using replications, the average performance of an entry over a sample of the environmental variation is obtained and aids greatly in generating a more valid evaluation. The second technique, called randomization, involves the random sequence of entry placement within each replication. Again, this is done to minimize the effect of a unique environmental condition in the experiment. Randomized sequences can be obtained from prepared tables of random numbers or by such simple techniques as drawing numbered slips from a hat or numbered beans from a jar. The only requirement is the complete impartiality in entry placement within replications. Figure 8.2 illustrates a random complete block experiment with four entries. Figure 8.3 shows a typical replicated field experiment plot layout to measure seed yield differences between orchardgrass varieties.

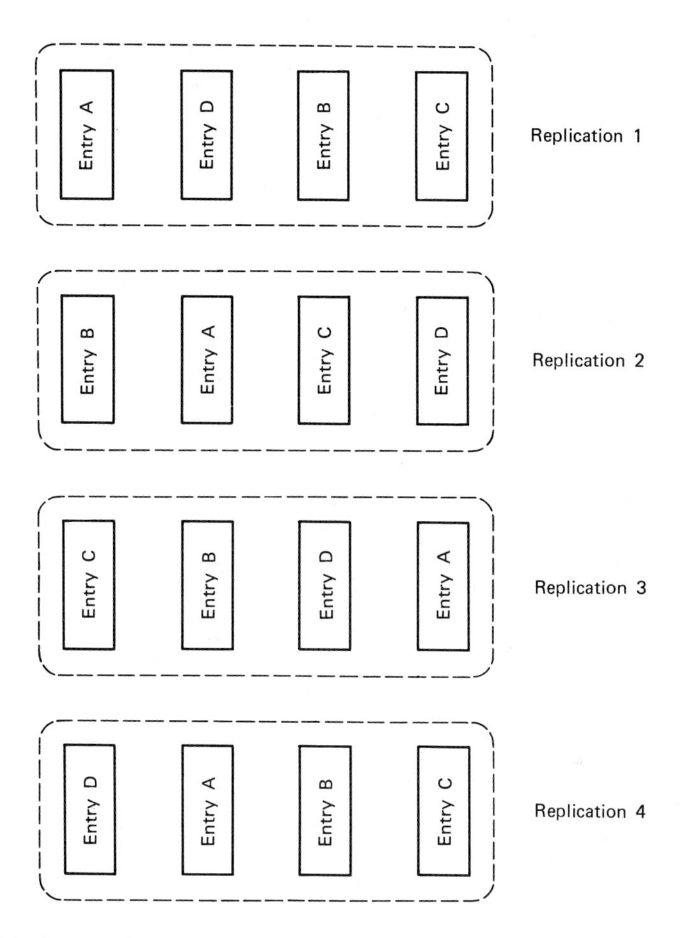

Figure 8.2. Four entries in a randomized complete block design with four replications.

In any type of experimental design, the analysis itself is based on average values and deviations from these values. From Figure 8.2 we can calculate an average value for the total experiment, another set of averages for each of the entries, and a third set of averages for each of the replications. By analyzing the amount of variance (deviations) from the average values for each source of variation, it is possible to calculate the relative amount of variation in the experiment due to the differences among entries, differences among replications, and experimental error. Error vari-

Figure 8.3. An orchardgrass seed yield experiment. The plots are four rows wide, five meters long, and replicated several times. The center two rows in each plot will be trimmed to remove border effects and harvested for yield. (Courtesy C. M. Rinckner, USDA-SEA.)

ance is so named because it remains unexplained in the experiment and may be an indication of accuracy levels in conducting the experiment. Following analysis, we can then assign probabilities that apparent differences among the entries are, in fact, real. We can use techniques to separate the entries themselves if the differences are judged to be real. The validity of any experiment is judged on the proportion of error variation compared with that for entries and replications. A very high proportion of unexplained variation will likely render the experiment invalid or at best very questionable.

Any experiment is designed to minimize the environmental effects that cause error in interpretation. With our apple trees at the beginning of the chapter, several things can be pointed out regarding field experimentation. Without question, the tree at the edge of the field was in a different environment than the tree in the middle of the orchard. Field edges generally represent "abnormal" environments compared with the total field area. Border effects that describe the environmental response of plants to either their neighbors or open spaces around them can seriously compound results. In our orchard it would be better to evaluate the trees surrounded by the same genotype than those having either empty spaces or abnormal phenotypic variations in the neighboring trees.

A common practice is to seed extra numbers of rows in each plot and then harvest only the center rows, leaving the border effect in the field with the unharvested rows. In this manner the variation from unnatural environmental interactions is minimized. Other things such as the uniformity of soil fertility, slope, and drainage are equally important. It is sometimes possible to evaluate geographic area with some uniform genotype prior to experimental use. This can be very helpful in making decisions on location of the experiment and the selection of the most appropriate experimental design. Many times, however, pre-assessed information on plot sites is not available.

By increasing plot size or replication numbers, or both, accuracy can be increased. Here a compromise must be made between highly precise test data and the amount of material to be evaluated. As plot size and replication numbers increase, the total number of entries that can be evaluated decreases. Generally, three to five replications are considered standard. Plot sizes vary greatly depending on the plant itself and the number of entries to be evaluated.

Many variations exist in experimental design and analysis. The design choice depends on the number of entries to be evaluated, usable space available, and available financial resources. Texts by LeClerg et al., (3), Snedecor and Cochran (5), Little and Hills (4), and Finney (2) adequately cover these subjects.

Association Evaluations

In many instances, the breeder is interested in the relationships that may exist between or among characters. Do the values of both characters change together and, if so, do they increase together, decrease together, or go in opposite directions? Important relationship questions may include: "Can yield and protein be increased together? What is the relationship between leaf size or shape and drought resistance?"

To evaluate relationships, correlation analyses are used in which values of two characters are analyzed on a paired basis. This means that the values for both characters on the same individual or plot are compared for association relationships. Figure 8.4 gives a generalized form of the three alternative relationship patterns. In the positive correlation relationship each character

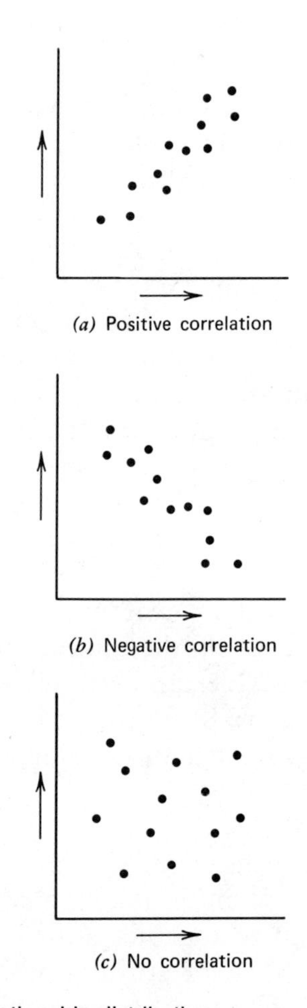

Figure 8.4. Generalized relationship distributions.

moves in an increasing direction. In a negative correlation the value of one character increases while the other decreases. Where no relationship occurs, the distribution pattern is completely at random. Note that each data point will contain both an environmental and genetic component. The relative magnitude of each component in the phenotypic value can vary, depending on the trait that is being measured and the environment in which it is evaluated.

Again, statistical analysis is required to determine the validity of the numerical relationship. In addition, if a relationship does exist, it may not be on a straight-line system but instead may follow some type of curve or other mathematical function. The establishment of a positive or negative relationship does not lead to a direct cause-effect interpretation. Careful consideration of the characters on a logical basis, and the application of proper statistical treatments, are necessary to correctly interpret the situation. Previously cited statistical texts are helpful in further study.

COMPUTER APPLICATIONS

Calculator analysis of data, until the late 1960s was normally a routine operation conducted in the offseason, resulting in a serious time lag between data collection and decision making in the breeding program regarding genotype selection and advancement. Producers and others who base operational decisions on experimental results were also seriously hampered by the long time periods between experiment completion and data presentation. The recent expansion of computer science and technology has resulted in a virtually unlimited set of possibilities for high speed analyses, summarization, and data storage. Figure 8.5 shows a typical computerized data collection sheet and analysis printout.

Most breeding programs now own or have access to computer facilities and use them extensively in recordkeeping and data interpretation. The net result has been greatly increased efficiency, much better use of time, and improved mathematical accuracy. Computer applications in breeding programs are covered in more detail in Chapter 9.

DRYLAND WINTER WHEAT VARIETY TEST

YEAR 74 | EXPERIMENT NO. 1 | LOCATION NO. 3 SPRINGFIELD

VARIETY NUMBER	VARIETY NAME	TR 1	TR 2	TR 3	REP	PLOT NO.	EMERGENCE DATE		HEADING DATE	MATURITY DATE	PLANT HEIGHT (IN.)	SHEAF WEIGHT G.M.S.	GRAIN MOISTURE	YIELD (GR.)	TEST WEIGHT (LBS/BU)
CI 13546	SCOUT				1	1		2	129	163	20			835	560
CI 11952	WICHITA				1	2		2	127	161	21			630	573
CI 13190	WARRIOR				1	3		1	132	168	18			815	571
CI 13999	TRAPPER				1	4		1	133	169	19			885	565
CI 13547	LANCER				1	5		1	131	167	19			700	574
CI 15075	CENTURK				1	6		1	131	166	17			710	550
CI 13996	SCOUT 66				1	7		1	129	169	19			740	562
CI 15891	BACA				1	8		1	128	162	18			740	557
CO723192	SN/TZP/Y54//TPR/3/043				1	9		2	131	166	18			700	575
CO723441	PT/Y/T/N//SUT/LCR				1	10		1	125	159	18			550	554
CO724072	II18889/TPR//CO652643				1	11		2	128	161	19			725	543
CO724085	II18889/TPR//CO652643				1	12		2	129	162	18			721	533
CO724091	II18889/TPR//CO652643				1	13		2	128	162	17			670	542
CO724101	II18889/TPR//CO652643				1	14		3	128	162	17			475	551
CO724377	II21031/TPR//CO652643				1	15		4	129	166	17			470	576
CO725049	II21183/2643//LCR/3/KS62				1	16		1	129	165	17			810	560
CO725052	II21183/2643//LCR/3/KS62				1	17		2	130	166	17			755	577
CO725055	II21183/2643//LCR/3/KS62				1	18		1	131	166	17			725	571
CO725061	II21183/2643//LCR/3/KS62				1	19		1	130	164	18			800	575
CO725082	II21183/2643//LCR/3/KS62				1	20		2	131	167	17			820	561
CG 2	CARGILL 501				1	21		2	132	169	16			915	564
CG 3	CARGILL 4				1	22		3	132	168	18			690	575
KS 70179	AGENT/4*SCOUT				1	23		1	129	162	18			875	569
CI 15068	EAGLE				1	24		1	128	162	18			995	572
CI 15068	EAGLE				2	31		1	128	161	18			725	
CO725055	II21183/2643//LCR/3/KS62				2	32		2	129	162	17			505	
KS 70179	AGENT/4*SCOUT				2	33		2	130	164	18			690	
CO725061	II21183/2643//LCR/3/KS62				2	34		2	130	164	17			720	

(a)

Figure 8.5. Using the computer in data collection and presentation. (*a*) A field book printed by the computer on special forms provided by the breeding program. Data are recorded in the form as they become available during the season. At the end of the year the information is entered into the computer for analysis and presentation. (*b*) A computer prepared and printed table of a four-replication experiment. The table is arranged in descending order of average yield for each entry. A statistical analysis summary is printed on the bottom of the table.

SUMMARY AND COMMENTS

The breeder will always be faced with the need to identify and quantify the sources of variability and thus obtain a much more accurate picture of the genetic gain to expect. Genetic variation is of primary interest in plant

TABLE WINTER WHEAT VARIETY TEST PAGE 1
1974 EXPERIMENT 1 LOCATION 3, SPRINGFLD WHEAT WINTER
SEEDING DATE--SEPT 22 HARVEST DATE--JUNE 24

5.44 INCHES PRECIP MARCH 1 TO HARVEST
15 PERCENT HAIL DAMAGE JUNE ? VARIETIES TENDING TO SHATTER SEVERLY DAMAGED

VARIETY NUMBER	VARIETY NAME	GRAIN YIELD	SPR SURV	HEAD DATE	MATUR DATE	PLANT HT.	TEST WT.
CI 13190	WARRIOR	19.7	1.0	132.	167.	18.7	57.1
CO725049	II21183/2643//LCR/3/KS62	18.8	1.3	128.	162.	17.7	56.0
CO725082	II21183/2643//LCR/3/KS62	18.7	1.5	129.	164.	17.7	56.1
CI 13999	TRAPPER	18.1	1.0	133.	168.	18.7	56.5
CI 15068	EAGLE	17.8	1.0	128.	162.	19.0	57.2
CI 13546	SCOUT	17.8	1.5	128.	162.	20.7	56.0
CG 2	CARGILL 501	17.7	2.0	130.	165.	16.7	56.4
CO725061	II21183/2643//LCR/3/KS62	17.6	1.5	130.	164.	18.0	57.5
CI 13996	SCOUT 66	17.2	1.3	129.	161.	20.2	56.2
CO723192	SN/TZP/Y54//TPR/3/043	16.6	1.5	131.	165.	19.2	57.5
CO724072	II18889/TPR//CO652643	16.3	2.3	128.	162.	17.2	54.3
CI 13547	LANCER	16.3	1.3	131.	165.	19.5	57.4
CI 15075	CENTURK	16.3	1.0	131.	165.	17.2	55.0
CI 15891	BACA	16.2	1.3	129.	162.	19.2	55.7
CO725052	II21183/2643//LCR/3/KS62	16.0	1.8	130.	164.	16.7	57.7
CO724085	II18889/TPR//CO652643	15.9	2.3	128.	161.	18.2	53.3
KS 70179	SAGE	15.1	1.8	130.	163.	19.0	56.9
CO724091	II18889/TPR//CO652643	15.0	1.8	128.	162.	17.2	54.2
CO725055	II21183/2643//LCR/3/KS62	14.9	1.3	130.	164.	17.7	57.1
CG 3	CARGILL 4	14.2	2.8	131.	167.	17.7	57.5
CO723441	FT/Y/T/N//SUT/LCR	14.1	1.3	125.	159.	18.0	55.4
CI 11952	WICHITA	13.1	1.3	127.	160.	21.0	57.3
CO724101	II18889/TPR//CO652643	11.8	2.8	129.	162.	17.2	55.1
CO724377	II21031/TPR//CO652643	11.5	3.8	127.	163.	16.5	57.6
GRAND	COLUMN MEAN	16.1	1.!	129.	163.	18.3	56.3

COLUMN	GRAND MEAN	S.E. PLOT	S.E. MEAN	S.E. DIFF	COEF VAR	F RATIO	TUKEYS (
GRAIN YIELD	16.13	2.1306	1.0653	1.5066	13.211	3.9900 **	5.58

(b)

breeding and a good understanding of statistical concepts is necessary to maximize efficiency in achieving breeding gains.

Most of my career has been spent in highly variable dryland environments with the associated problems of wide temperature fluctuations, unreliable rainfall patterns, and hail storms just prior to the harvest season. The challenges of this kind of environment are often discouraging, and yet, when success is achieved, the rewards and satisfaction are great indeed. This environment stimulates humility because just about the time one is confident of the answers, the questions are changed. Breeding will always represent a battle to improve the genetics of the plant under conditions of environmental variation.

I am continually amazed with the potential range of breeding application possibilities in computer technology. Contemporary students are normally skilled in computer systems and quickly adapt new procedures for breeding program use. I feel that it is a waste of time to do anything manually that can be done electronically yet my only concern is that we maintain enough manual operations so that concepts are learned and understood. Beyond that, let the machines do the work!

REFERENCES

1. Allard, R. W. 1960. *Principles of plant breeding.* Wiley, New York.

2. Finney, D. J. 1972. *An introduction to statistical science in agriculture.* Wiley, New York.

3. LeClerg, E. L., W. H. Leonard, and A. G. Clark. 1962. *Field plot technique.* Burgess, Minneapolis.

4. Little, T. M., and F. J. Hills. 1978. *Agricultural experimentation.* Wiley, New York.

5. Snedecor, G. W., and W. G. Cochran. 1967. *Statistical methods.* 6th Edition. Iowa State Univ. Press, Ames.

9

BREEDING OBJECTIVES

Every breeding program has a set of stated or implied objectives. The number and interrelationship of these objectives depends on the plant species and the environment in which the plant is to be produced. The importance of establishing a clear-cut set of objectives relates to efficient use of the resources (primarily financial) available to the breeding program. Well-defined objectives, for example, may involve the use of a single experimental plot in several simultaneous investigations, which results in increased information output with minimum input. The objectives must be clearly stated so that the breeder is constantly aware of the most important priorities in the breeding effort. Poorly defined objectives lead to the unwise use of limited financial resources through duplication of effort, uncompleted research, and misconceived breeding priorities.

The breeder must have a thorough knowledge of the method by which the species under improvement will be reproduced and placed in the hands of the growers. General categories include pure lines, hybrids, synthetics and composites, and asexual reproduction. As we will see in specific chapters on breeding systems, the methods of improvement overlap extensively between and among reproduction categories. In addition, the reproduction systems are not mutually exclusive. For example, it is entirely possible to have vegetatively reproduced hybrids in the ornamental species. It is important to be aware of all genetic improvement techniques that can be employed for any species, and to understand the limits that might be imposed by the reproduction system.

The relative amounts of genetic and environmental control on the character or characters under investigation will affect the implementation and achievement of plant breeding objectives. If virtually all the variation in the character is environmentally controlled, then a majority of the program resources and effort could be directed toward unrealistic objectives. On the other hand, a character that is primarily genetically controlled presents a readily achievable goal, providing adequate genetic variability is available. This statement does not imply that we should breed only for highly heritable traits. To do so would result in the avoidance of many economically important plant characteristics with low heritabilities. The breeder should, however, be aware when objectives involve low heritability characters to properly adjust program design and thereby maximize efficiency.

A few common large magnitude breeding objectives are discussed. Many additional program objectives will become evident as breeding technique chapters are presented.

YIELD

Yield improvement is the ultimate goal in virtually every plant breeding program. The units of yield depend entirely on the plant species under consideration. It might be expressed in terms of tons/hectare in food or feed crops, or number of blooms per plant for an ornamental species. Often, but not always, we are looking for an increased quantity of a directly marketable product. Yield generally can be quantified, and is frequently related very closely to another major objective, quality, which is discussed in a following section.

The character of yield reflects the performance of all plant components and might be considered as the final result of many others. Improved yield can generally be attributed to two major causes. First, every plant contains an inherent physiological production capacity that operates on energy, nutrients, water, and other natural resources required for normal plant performance. All genotypes do not have the same inherent physiological capacity to yield. In rice, for example, the introduction of semi-dwarf genes reduced plant height and improved the plant's capacity to more efficiently use the natural resources at its disposal. Some of the increased yield potential was due to the stiff straw characteristics of the semi-dwarf that allowed it to stand erect and maximize use of very high irrigation and fertilizer levels. However, some improvement was also the result of other genetic information leading to a greater inherent yield potential through improved components such as numbers of grains per panicle. Breeders commonly find yield to be a very complex array of plant component interactions, and by the manipulation of these genetic systems yield is improved as the result of plant efficiency improvement. The second major area of potential yield improvement is the protection against environmental hazards, such as diseases and insects, drought, winter injury, and high salt levels in the soil or irrigation water. These represent separate breeding objectives and are discussed in subsequent sections of this chapter.

While plant responses that make up improved yield can often be identified, we see very quickly that yield is a highly complex trait that may be the end result of many interacting factors. The heritability of yield is normally low because of the large number of genes involved and the high level of environmental interaction. With very difficult environments, yield improvement through breeding is achieved through small increments over long time periods.

Despite all the difficulties associated with yield, improvement in this character has been achieved through breeding and genetic techniques in virtually every species studied. Some of the most dramatic yield improvements on record have been realized in controlled environments with few limitations imposed on the plant. The "Green Revolution" varieties of wheat and rice are good examples. Ishizuka (8) in summarizing rice yield improvement showed that a doubling of yield from 1 to 2 tons/hectare was accomplished in a one thousand year period from 900 A.D. to 1885 but another doubling from 2 to 4 tons was achieved in the next period of less than one hundred years. He attributes the recent rapid rate of improvement to a combination of the following factors:

1. New varieties developed by scientific and systematic breeding
2. Improvement in cultivation techniques
3. Chemical fertilizers, especially nitrogen
4. Fungicides, insecticides, and herbicides
5. Soil improvement

As he points out, yield is a package deal in which the genotype is dependent on environmental production factors to experience maximum expression. In modern research programs, yield improvement is generally obtained through the team effort of several scientific disciplines.

QUALITY

Quality is a component that adds value to the crop. As with yield, the definition of quality depends on the plant and its intended use. Quality may mean the nutritional value of grain to a cattle feeder, or taste and texture of a tomato to a gardener. Regardless of the definition, quality is an important component of any breeding program.

Consumed Crops

Many crops are produced for either animal or human consumption and each group has certain sets of quality properties. In crops for human consumption quality will include nutritional and nonnutritional properties. Nonnutritional properties include physical and taste characteristics. Examples include quality of beer produced from the malt of barley varieties, and the color and flavor of different apple varieties. These properties can have very important economic implications. For example, the wine quality produced from different grape varieties with specific environmental interactions for each location and year can add value to the product very dramatically. Some wines command a high premium on the market because of the correct combination of variety and environment. We also know that there are distinct baking property differences among wheat variety flours with respect to the size of loaf, the grain and texture of the bread, and the color of the

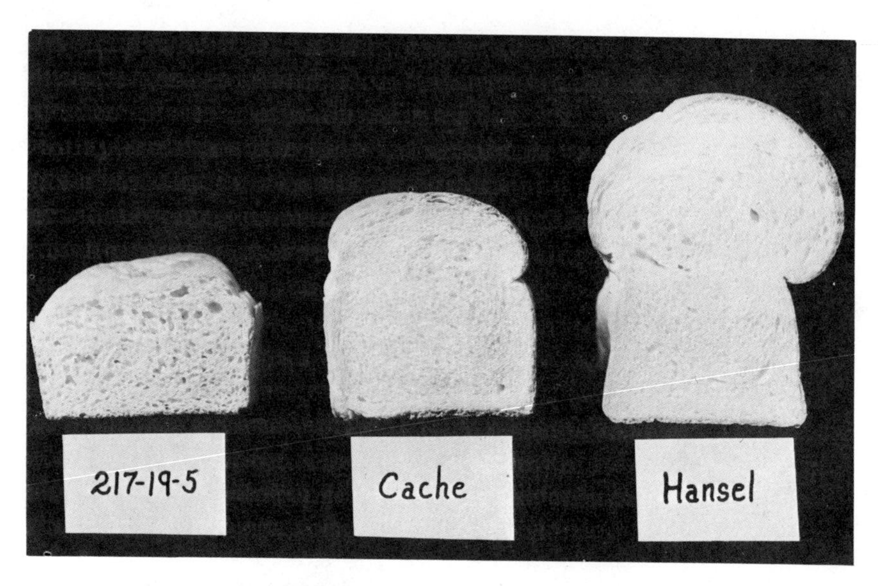

Figure 9.1. Bread baking quality differences in wheat. The loaves have been produced from different varieties with the same protein levels. Each loaf started with the same amount of ingredients. The loaf sizes, shapes, and internal properties are the reflection of the different quality genotypes. (Courtesy W. Dewey, Utah State Univ.)

crust (Fig. 9.1). However, flour properties are also a reflection of the environment in which they were produced. Thus, each of these product properties is under some level of genetic control but environment also contributes to the final character expression.

Every society has its own sets of product standards. In grains, for instance, such items as color and presence or absence of hulls will have different relative values, depending on how the product is used. Product properties that do not have society acceptance often lead to the abandonment of otherwise excellent genotypes.

Genetic variability has been demonstrated for most product properties and progress is possible through selection in a breeding program. On the other hand, virtually all product property characteristics have a medium to high level of interaction with environment, meaning that they must be evaluated in several environments before a true average performance picture can be obtained.

In recent years much research has been devoted to the area of nutritional properties for human consumption. Commonly used human consumption foods, especially the major grain crops of wheat, rice, corn, barley and sorghum, are deficient in one or more of the essential nutrients for normal human growth and development. Special emphasis has been placed on increasing the content of lysine and some of the other limiting amino acids.

The first breakthrough in genetic control of nutritional properties took place in corn with the discovery of the opaque-2 (o_2) gene that improved lysine content. This gene not only changes the amino acid balance but also alters other endosperm characteristics including kernel development and density. Other examples of crops with genetically controlled nutritional quality include sorghum, wheat, barley, and rice. In each of these crops genetic variability has been demonstrated for both protein content and amino acid balance. Since cereals are included as a high proportion of the human diet in much of the world, genetic improvement of nutritional properties will result in improved human diets, an especially important consideration in developing countries with subsistence agriculture. The improvement of nutritional components through genetic control has been the subject of a number of conferences and workshops and is summarized in a 1976 National Academy of Science workshop (6).

Human nutritional properties in crops other than those produced for seed are also under genetic control. For example, genetic variability exists for the vitamin C content in strawberries and has been improved through breeding. In many of the fruits the normal levels of nutritional properties are high enough so that extensive effort has not been warranted to improve these characteristics.

Quality properties such as palatability and nutrition in forage crops can be altered genetically to improve the acceptability and value of the crops for animal use. Burton (3) summarizes the potential for breeding better quality forage plants. Several types of improvement are possible. The introduction of dwarfing genes in pearl millet increases the leaf percentage in the forage crop, which, in turn, increases the consumption rate dramatically. By the manipulation of photoperiod response—a genetically controlled trait—maturity can be changed in several forage crops so that better quality hay is produced. Some forages contain anti-quality chemical components such as coumarin that block or inhibit proper metabolism. These can be reduced genetically, thus improving the quality properties. As with most other quality considerations, these characters are subject to extensive environmental variation.

Aesthetic Crops

Ornamental plants, bred and produced for their aesthetic properties, also have a wide array of quality characteristics. These include the length of bloom time, flower and foilage color properties, desirability and intensity of fragrance, and flower shape. In roses, for instance, some quality objectives include stability and clearness of colors, fragrance, thornlessness, and high petal numbers. Gorer (7) lists the characteristics and development of a number of garden flowers. Because of the wide array and diversity of these species, the breeder who is interested in improved ornamental plant quality must become thoroughly familiar with the plant being considered.

Storage Properties

Storage properties of perishable crops are another genetic consideration in quality. Crops with storage quality problems include fresh fruits and vegetables and potatos. Genetic variability is demonstrated by some varieties of potatos that store much better than others. This may result in part from the genetic control of storage diseases but may also be due to improvement in inherent biochemical storage properties of the variety. The physical environment including temperature, relative humidity, and light will have significant effects on and can interact with the genotype in the storage of the crops. Storage quality has significant economic implications, as storage problems can change a crop from a profit to loss situation very rapidly.

PEST RESISTANCE

Almost without exception, every plant is exposed sometime during its life cycle to an array of biological pests such as weeds, insects, and diseases. The extent of the damage and crop loss,

Figure 9.2. Disease losses in corn. The row on the left has been severely infected with virus while the row on the right is disease free. (Courtesy W. K. Findley, USDA-SEA.)

which can range from almost an unmeasurable quantity to a disaster, depends on the combination of host genotype, pest genotype, and environment (Fig. 9.2). Often those environmental conditions conducive to high crop yields also contribute to maximum pest development. For example, moisture levels adequate to produce heavy plant canopy and high foilage density also lead to high relative humidity within the plant community, which in turn encourages the development of insect populations and diseases. This statement is subject to many exceptions. Vigorous crop growth will generally provide good competition to keep weed growth to a minimum. Occasionally a disease or insect attack will occur following the weakening of the plants by some severe environmental stress such as drought. It is, therefore, very important to understand the epidemiology of the pest and its interaction with the host.

Pest resistance is vital in the production of maximum yield. If a plant has extremely high inherent capacity to yield but lacks pest resistance mechanisms, then maximum production cannot be realized if a pest attack occurs. The development of high quality products in food, feed, and aesthetic crops can easily be impaired by untimely pest attacks. Storage properties are also seriously affected by various pest problems. Pest resistance thus becomes a significant part of all breeding programs.

Pest Biology

All pest organisms contain mutable genetic systems. Recombination can occur provided some sort of sexual or nuclear fusion mechanism is available. Without genetic variability in the pest, resistance breeding would be an easy task since a single form of resistance in any plant would last indefinitely. This is not the case, however. As the genetic information in the pest changes because of mutations and recombination, different races or biotypes develop that have new capacities to attack the host organism. The concept of natural selection is important in plant pest populations. In general, if large acreages of a highly resistant crop are planted, natural selection is applied to the pest so that any potential genetic change that could result in virulence (the ability to attack) has a high selective advantage. The result is an increase in the population of pest organisms that can again attack the host species. Because of this potential for variability and natural selection, pest resistance breeding programs are a continuing operation with acceptable levels of resistance or tolerance being only temporary.

Host-Pest Interactions

An interesting concept in host-pest relationships was developed by Flor (5) in his research with flax and flax rust. He showed that gene mechanisms operated in both the host and pathogen, and resistance or susceptibility resulted from specific combinations of corresponding alleles in both organisms. This basic concept, called the gene-for-gene hypothesis, clearly illustrates an intimate relationship between the host and the pest. Following Flor's work, the refinement of host-parasite mechanisms has been extended to many crops.

There are two general categories of resistance mechanisms called specific or vertical, and general or horizontal. In specific resistance systems a single gene with major effects conditions the resistance to the invading organism. Specific resistance can also be considered as a very narrow range of protection to the array of genetic variability in the pest. Based on the gene-for-gene hypothesis, a single allelic change in the pest attack mechanism can result in the breakdown of specific resistance. This type of resistance generally results, however, in a very high level of resistance or immunity until the host-pest genetic relationship undergoes changes.

General resistance is conditioned in the host by a large number of loci with alleles acting in an additive or complementary manner. This results in a stable protection mechanism not highly subject to radical change when a single mutation event in the pathogen occurs. The host has protection against a wide array of genetic variability in the pest. General resistance is expressed as a reduced compatibility between the pest organism and the host, and a low level of infection may be displayed by the host, making phenotypic selection difficult. Some levels of the disease may be present in the crop during the year, but losses are greatly reduced. Because of the natural selection concept, general resistance is often proposed to be more permanent than specific resistance since the pest organism is allowed to maintain a population in limited numbers, thus reducing intensive selection pressure for highly virulent types. General resistance is still under considerable investigation to determine its exact nature and proper methods of selection.

A wide array of resistance or avoidance mechanisms are displayed by different plant species. In some systems the host produces a chemical compound that is detrimental to the health and reproduction of the pest. In beans, for example, a compound called phytoalexin is produced by the host in response to infection by the disease, *Fusarium* wilt, which then inhibits the growth and development of the disease. The presence of morphological structures such as pubesence may prohibit the pest from attacking the plant as, for example, with cereal leaf beetle resistance in wheat. The hairy leaves deter the female from laying eggs and thus interrupt the normal reproductive pattern of the pest. Many variations in host-parasite mechanisms can be found in nature.

Resistance Breeding

Resistance breeding in a plant improvement program requires several considerations. First, the level of pest damage must be determined to make priority assessment possible. The economic impact of some plant pests is well documented. With others, relatively little is known about the disease, and baseline data including crop loss estimates must be generated. Effectiveness and permanency of other control methods must be evaluated.

If resistance breeding is judged to be a major goal, then the amount of genetic resistance available to the breeding program must be determined. Again, the state of information may depend on the pest under consideration. Screening of a large amount of material, including world collections and related species, may be necessary to identify usable resistance genes. If natural variability is difficult to find or utilize, artificially generated mutations may be a reasonable alternative.

Following the evaluation of the variability, the resistance is incorporated through breeding in combination with other characteristics important for production and quality. Breeding approaches and techniques are discussed in subsequent chapters.

Finally, continual monitoring of the pest system is necessary to be fully aware of any biological changes taking place. This is often accomplished by regional disease laboratories operated by the federal government. Through the use of well-defined standard host varieties and genotypes, races or biotypes of the pest that exist during any growing season are identified following collection from throughout the growing region.

A final problem is associated with the permanence of resistance in different crops. In an annual crop such as the cereals, new resistance can be generated and incorporated in a few years. Since the crop is replanted each year, resistant varieties can rapidly be placed in production. In crops of long cycle, such as fruit tree species, the time required to put new resistance into production makes immediate response disease breeding very difficult.

Nelson (9) and Russell (12) provide detailed information on pest resistance mechanisms in many of the commercially important agronomic crops. Painter (10) and Russell (13) discuss insect resistance in crop plants. Resistance breeding in fruits is covered in "Advances in Fruit Breeding" (1). A monograph edited by Maxwell and Jennings (2) provides a comprehensive review of insect resistance breeding.

Pest outbreaks are highly visible to the general public and can be a driving force in the development of a breeding program. It is easier to generate financial support because of disease or insect-caused losses than for almost any other plant breeding goal. After yield, pest resistance breeding is the single most common goal in all plant breeding efforts. It is also the most economical in producing returns for the grower. Genetic resistance to plant pests offers a built-in insurance policy with negligible out-of-pocket cost to the grower at the time of seed purchase, and provides control without the complications of chemical additives in the environment, associated mechanical costs, and application time.

IMPROVED ADAPTATION

All plants have adaptive mechanisms allowing them to exist in a complementary manner with the environment. Wilsie (14) describes the environmental parameters that dictate the ecological habitats for many crop species. Factors interacting with the physiological mechanisms of the plant in determining adaptation include temperature, photoperiod, wind, and moisture. These major items with their array of fluctuations provide a set of controls for plant limits in viability and production. Plant breeders consider the genetic response to environment as a group of adaptation genes necessary for satisfactory plant production.

Domestication

Domestication as a form of evolution was discussed in Chapter 7. This is a continual process that has been utilized to mold plant species to human needs. Much of the domestication occurred early in human society evolution, but even today new crops are being considered. Examples include wildrice for cereal consumption and jojoba, a desert plant, for high quality oil production.

Domestication in modern production systems requires selection for specific traits with accompanying reduction in genetic variability. This is turn leads to increased genetic vulnerability. As domestication proceeds, the breeder must continually evaluate the priority of goals in the context of conventional breeding considerations such as yield, quality, and pest resistance. As in all other breeding programs the potential value of the species has to be evaluated before a long-term domestication program is undertaken.

Improved Adaptation

Often, plants are produced in geographic areas with environmental conditions for which they may not be particularly well adapted. New crop species introduced during the settlement of the United States are a good example of plant adaptation. After several false starts with spring habit wheat and corn, certain areas of the Great Plains were identified as having a high production potential for winter wheat. This is logical, considering that winter wheat was introduced from comparable environments in Europe and Asia. However, of the introduced varieties, only a few have been successful because of the rigorous climate in this geographic area and the poor adaptive levels of many genotypes.

Two approaches can be taken in the improvement of plant adaptation. First, the environment can be altered in such a way that the plant is not subjected to abnormal stresses. The development of the deep-furrow drill permitted the production of winter wheat in South Dakota and Montana where the environment has previously been too harsh for plant survival. The environment was altered so that more winter protection from dessication and cold was provided for the plant allowing it to survive satisfactorily where it normally would have died.

The second approach is to alter physiological mechanisms associated with adaptation. A classic example is sorghum where the maturity responses were shortened genetically to allow production in a much more northernly latitude than had been possible with introduced genotypes. In this case breeders altered the physiological mechanisms under genetic control to change the adaptive system of the plant.

ENVIRONMENTAL STRESS TOLERANCE

Environmental stresses, as distinguished from plant pest problems, include drought from high evapotransportation (Fig. 9.3), cold stress either in the form of freezing or low temperatures not conducive to normal plant function, high salt levels, and air and water pollution. Much of the world's food production takes place in regions where the crop must survive and produce while being subjected to the whims of nature. Improvements in genotypes and cultural practices aid in stabilizing plant performance but the stresses will always be present.

Environmental stress problems are generally very difficult to deal with by breeding because their complex nature requires a wide array of genetic response mechanisms. The plant response to drought, for example, is not well understood physiologically but is the net result of several systems including the coating on leaf surfaces, the number, size, and response of stomata and the ability of the root system to function efficiently during periods of moisture deficiency.

Despite the difficulties, geneticists and plant breeders have accepted improved response to stresses as achievable plant breed-

Figure 9.3. Drought stress in corn. Major symptoms include leaf rolling and reduced foliage development. (Courtesy W. K. Findley, USDA-SEA.)

ing goals. For example, in 1976 a workshop on plant adaptation included a section on the genetic potential for solving problems of soil mineral stress (11). Screening techniques usable for selecting genotypes with acceptable levels of tolerance to such salts as aluminum, manganese, and iron were discussed. This is only one example of the current interest in developing genetic potential through breeding for improved plant efficiency under environmental stress conditions. The publication "Climate and Food" (3) also addresses the question of genetic variability for adaptation under climatic stress conditions.

MECHANIZATION OF CROP PRODUCTION

With rising production costs, producers are continually seeking ways to produce high value crops more efficiently. Genetic variability has been demonstrated for a number of cost-saving traits.

In sugar beets, many seeds are normally produced in one seed ball. When planted in a beet production field, the rows must be thinned, often by hand, to the desired plant density. A gene was discovered that converted the plant from multigerm to one seed per seed ball (monogerm). Seed from monogerm plants can be precision planted, greatly reducing thinning costs. The monogerm gene has been used extensively in beet breeding programs and nearly all beet varieties now have this character.

Mechanical harvesting equipment has been developed or is being considered for several fruit and vegetable crops. Traits that can be altered by breeding include constant fruit size, uniform ripening, resistance to mechanical injury, and ease of fruit removal from the plant. Mechanical harvesting has been developed for tomatoes, and is being seriously considered or tested for other crops such as cucumbers, grapes, apples, and pears.

SUMMARY AND COMMENTS

Any breeding program, if effective, will have a well-defined set of goals and objectives. They must be appropriate to the plant problems and have a reasonable likelihood of being achieved. Some breeding objectives such as disease resistance may achieve rather dramatic results in a short time. Others, such as plant adaptation, may represent slow but gradual improvement. A breeder who thoroughly understands the objectives will invariably develop a good base of background knowledge. Without exception, both genotype and environment influence the plant response. Often the accomplishment of breeding objectives requires the input of more than one scientific area.

In my own experience I have always been faced with many more potential objectives than I could realistically pursue. Because of the limited water resources in the dryland production system, my objectives were primarily those of adaptation rather than plant pests, although the presence of pests has occurred with enough frequency to make me appreciate the challenges of both. My best source of information and guidance regarding the establishment of breeding objectives has always been the producer. Because he or she is faced with the problems of making the plant perform under a set of environmental conditions, they normally can provide excellent experience input into the establishment of the work that a breeder should be doing. I do not ignore my scientific colleagues, but do place great weight on the opinion of people who are on the firing line every day in their crop production activities.

REFERENCES

1. *Advances in fruit breeding.* 1975. J. Janick and J. N. Moore (eds.). Purdue Univ. Press, West Lafayette, Ind.

2. *Breeding plants resistant to insects.* 1980. F. G. Maxwell and P. R. Jennings (eds.). Wiley, New York.

3. Burton, G. W. 1978. *Advances in breeding a better quality forage plant.* Proc. 27th Annual N. C. Cattlemen's Conf. and 1978 Forage and Grassland Conf. pp. 96–102.

4. *Climate and food.* 1976. Natl. Acad. Sci.-Natl. Res. Coun., Washington.

5. Flor, H. H. 1956. The complementary genic systems in flax rust. *Adv. Genet.* 8:29–54.

6. *Genetic improvement of seed protein.* 1976. Natl. Acad. Sci.-Natl. Res. Coun., Washington.

7. Gorer, R. 1970. *The development of garden flowers.* Eyre and Spottiswoods, E. C.

8. Ishizuka, Y. 1969. Engineering for higher yields. pp. 15-25. In J. D. Eastin et al. (eds.). *Physiological aspects of crop yields.* Am. Soc. Agron., Crop Sci. Soc. Am., Madison, Wisc.

9. Nelson, R. R. 1973. *Breeding plants for disease resistance.* Penn. State Univ. Press, University Park.

10. Painter, R. H. 1968. *Insect resistance in crop plants.* Univ. Press of Kansas, Lawrence.

11. *Plant adaptation to mineral stress.* 1976. M. J. Wright (ed.). U. S. Agency Int. Dev., Washington.

12. Russell, G. E. 1978. *Plant breeding for pest and disease resistance.* Butterworth, Woburn, Mass.

13. Russell, W. A. 1975. Breeding and genetics in the control of insect pests. *Iowa State Jour. Res.* 49:527-551.

14. Wilsie, C. P. 1962. *Crop adaptation and distribution.* W. H. Freeman, San Francisco.

10

PROGRAM DESIGN AND MANAGEMENT

All breeding programs are intended to improve the usefulness of the plant. Improvement can be achieved in a number of ways, depending on the plant species, the environment, and the program goals.

One consideration is common to every breeding program. Time is required to develop a new hybrid or variety. This can range from a few to many years depending on the generation time. In annual plants, for instance, it is possible to use greenhouse and alternate growing site facilities to produce several generations per year. Seed may be harvested shortly after embryo formation without complete maturation. These types of techniques can reduce generation time significantly. In fruit tree crops, on the other hand, an extremely long period is required to produce each generation, even though the forcing of juvenile plants to flower and other innovations have helped reduce generation time. A minimum of 8 to 10 years is required to get a new variety or hybrid of any species into commercial production.

A plant breeding program is often compared to a pipeline where new genetic combinations are placed in the inlet each year, successively moved along from generation to generation with appropriate selection techniques, and new varieties or hybrids finally appear at the outlet. New productive genetic combinations must be entered into the pipeline each year with proper selection techniques applied to have a relatively constant output of new varieties. The total economic cost of the program is not born by each variety but is prorated over the varieties produced in a given period of time.

Because of the time element, support cost of a breeding program is generally large. Continuity over a period of years requires a sizable commitment of funds for time periods that are longer than normally considered in research programs. Any interruption in the pipeline flow chart results in erratic production of usable material that in turn severely reduces the system efficiency. Because of extensive financial commitment, maximum efficiency must be built into the design and management of the breeding program.

A breeder will either step into or initiate a program. In either case, management and design become extremely important in efficiently carrying out breeding objectives.

PROGRAM RESPONSIBILITIES AND FUNDING

In the U. S., plant breeding programs are funded by either public or private sources. In Western Europe most programs are conducted by private organizations, while in Eastern Europe and Asia practically all breeding effort is supported by government sources. Our discussion covers primarily the U. S. system. Most of the principles, however, could be translated to programs throughout the world.

Public Programs

Public plant genetic and breeding programs are supported either by the federal government through USDA or by a state experiment station system. USDA programs are normally affiliated with universities, although they may be located at independent research stations. Support funds are provided by federal legislation. Federal breeders will serve as associate members of university plant science departments and are often major professors or committee members for plant breeding and genetics graduate students.

Experiment station programs are integrated with the Land Grant system and breeders are members of collegiate departments such as agronomy, horticulture, and botany within the university. A project funding base is supplied through state tax money. Frequently, outside financial support is received from grower groups as well as contract and grant agencies. Grower support is normally directed toward specific applied breeding objectives, while contract and grant money more often is used in basic research problems. The breeder is required to develop grant proposals and project outlines for funding within the public system and for presentation to potential outside sources.

Efforts in the public program are divided into three general areas. Practical breeding responsibilities for the development of varieties or hybrids that will be usable by the public often occupy a major portion of the program effort. For instance, most of the cereal and fruit varieties in use today in the United States have been produced from state and federal breeding programs. In addition, germplasm not released as a named variety is often made available to private breeding organizations. For example, the sugar beet genetics and breeding group in the USDA Crops Research Laboratory at Fort Collins, Colorado, develops germplasm with important genetic properties such as disease resistance that is then made available to private breeding programs for use in producing commercial sugar beet varieties and hybrids. Often, the breeders with federal programs will have the responsibility of coordinating regional evaluations that include entries from a number of other state and federal programs. These regional nurseries are particularly helpful in evaluating germplasm over a wide array of environments. Project revisions, progress reports, and achievement statements are required periodically in both state and federal programs.

A second area of public program responsibility is the development of basic research information that includes such subjects as host-pathogen relationships, the inheritance of plant characteristics, the genetic control of plant metabolism, and the interaction of genotypes with cultural practices. This type of information, generally reported as scientific literature, has contributed extensively to our highly productive agricultural system.

A third area is the training of students for plant breeding careers. This aspect is tied very closely to research objectives, since the students normally conduct basic or applied research on the crop in the breeding project as part of their training. Student training includes the development of proper scientific investigation methods, the application of these methods to a research project, and the reporting of the subsequent results in thesis form and/or scientific publications. The student is also exposed to practical breeding concepts by working with the breeder.

Private Programs

In privately funded plant breeding, the underlying program motivation is to generate a profit for the company. Many commercial plant breeding programs have been extremely successful in contributing to high productivity levels in American agriculture. The success of any breeding program is dependent on the development of germplasm competitive with materials produced by other companies as judged by individual producer performance evaluation. Since producers must make a choice of hybrids and varieties in their production management, they serve to keep the competitive system operational in private plant breeding.

Private breeding programs are highly oriented toward applied breeding goals of variety and hybrid development. Private companies are most interested in plant species with legal or biological protection against others utilizing their material. Legal protection systems are discussed in Chapter 20. Biological protection for hybrids, supplied in the form of inbred line control, will be covered in Chapter 16.

Over the years public and private sectors of plant breeding have put forth considerable cooperative effort to develop germplasm and production systems leading to modern efficient agricultural production. Active variety and hybrid breeding programs remain with the public agencies until private industry can visualize profitable ventures in the crop. For example, practically all corn breeding effort was conducted by public agencies until the refinement of the hybrid corn system involving inbred lines. Since the development and acceptance of hybrid corn, virtually all hyrids are now sold privately. Public agencies still generate basic corn information but are generally not involved in the development of specific hybrids. They do, however, make inbred lines available to private companies for their use in hybrid breeding.

UNDERSTANDING THE PLANT

A good working knowledge of the plant is paramount to designing and managing an efficient program. A thorough knowledge of the reproductive system is of prime importance. Successful crossing to produce new genetic combinations is dependent on understanding the mating system. This may require additional research if information is not complete. In addition, knowing the reproductive system will aid in the interpretation of segregating population genetic composition, necessary to conduct efficient selection programs.

Another area of prime concern is the genetic variability available to the breeder for potential gain. The breeder must be aware of variability sources and potential problems of their incorporation into improved varieties. Difficulties may be encountered because of incompatibility mechanisms, problems in making interspecific crosses, linkage with unacceptable traits, or simply the lack of variability in the germplasm array available to the program. The genetic variation potential is a major factor in identifying realistic program objectives.

Common cultural practices employed in producing the plant must be well understood. Production hazards and problems will play a vital role in establishing breeding objectives. Currently used and potential future production systems should be considered. For example, the concept of no-till production practices where chemicals control most of the weeds is becoming more and more attractive, as production costs increase and additional pressure is applied to reduce wind and water pollution sources in agriculture. As this type of production system becomes common, a new array of genotype-environment interactions will occur. Diseases, insects, weed control, soil temperature, and moisture will take on different relationships from those in existing practices. The breeder must be aware of potential production changes so breeding objectives can be altered to fit new requirements. Generally the producer is an excellent source of information on current cultural practices but input should also be obtained from other plant production scientists.

BREEDING OBJECTIVES AND PRIORITIES

The identification of breeding objectives may not be particularly difficult, and those discussed in Chapter 9 are common to many programs. The difficulty lies in establishing priorities for objectives so the breeding program can be organized in an orderly productive manner. Breeders are generally a curious lot and can have difficulty in establishing and maintaining orderly program priorities. However, more problems exist than can be accomplished with existing financial resources and priorities must be established or the breeder becomes spread too thin and reduced program effectiveness occurs. This leads to frustration for the breeder and disappointment for the funding organization.

The ease of establishing objective priorities may depend on the age and experience of the breeder. Professionals who have been in the business for many years can often recognize priorities from past experience, and will have little difficulty in selecting those things that should most logically receive attention. A new breeder starting on an initial assignment may experience confusion and doubt when considering program objective priorities. The input of several people may be required. Scientists, growers, and administrators and management are commonly included. Scientists will provide background information about the plant production systems, plant hazards, genetic variability, and potential genetic gain for a number of objectives. Growers will provide practical experience, and help identify objectives most pertinent to their production systems. Administrators and management personnel are extremely important as they are responsible for providing financial resources. Complete understanding of and contribution to priorities leads to a high level of confidence and support.

IMPLEMENTATION

The collection of background information and establishment

of program objective priorities is followed by implementation of the breeding effort to efficiently reach the objectives. In aggressive programs, breeders are continually faced with more segregating material than they can feasibly evaluate and maximum efficiency techniques are imperative. The number of test sites is one of the most difficult, but important, decisions a breeder must make. As the number of locations increase the cost goes up. Since the breeder cannot test all environments, a minimum number must be selected to provide adequate information for accurate selection decisions. The choice may be reached after a series of preliminary investigations. Normally, a breeder will evaluate the environment for several years before the combination of sites is finally chosen. Even then, questions will exist about environments not being tested. The breeder must continually evaluate this decision and be ready to make changes if the situation appears to warrant it.

Statistical designs are varied and the choice may depend in part on amount and type of environmental variation present. Often, statistical design modifications such as the frequent inclusion of a check variety with a reduction in replication number may be employed. Many other statistical modifications and innovations are available.

Breeding programs are labor intensive because of small plant material or seed quantities and the need for individual breeding line assessment by manual operations. The ability and skill employed by the breeder in organizing the available labor force is critical to program efficiency. As labor costs increase, the development of mechanization is an opportunity to increase output with given financial resources. Breeding systems mechanization often originates with the breeders themselves. Small plot combines, seed packaging equipment, and electronic data collection are examples. Figure 10.1 to 10.3 illustrate types of mechanical innovations used in plant breeding programs. Breeding equipment is often not mass produced or marketed commercially because of low volume demand.

Computer applications aid significantly in the reduction of labor costs, but computerization may require the assistance of personnel with special skills such as programming. Once the basic programs are in place, however, minor modifications and updating

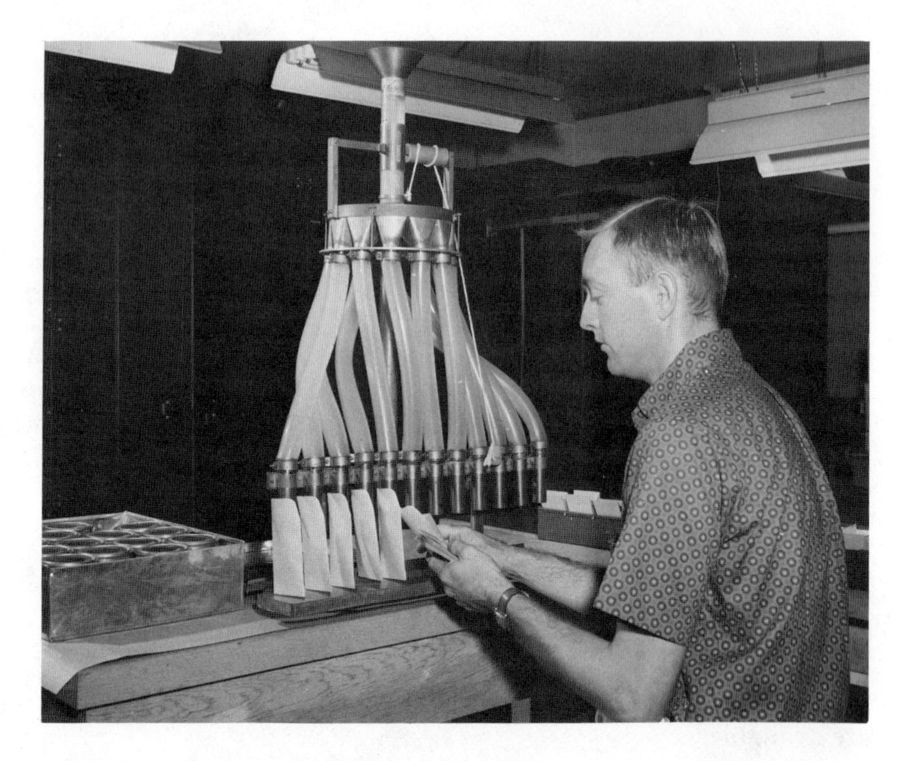

Figure 10.1. Mechanical seed packaging equipment. A known quantity of seed is placed in the top reservoir. When tripped it automatically divides the seed in equal volumes among the envelopes on the delivery tubes. (Courtesy J. C. Craddock, USDA-SEA.)

can generally be accomplished by personnel with a minimum of computer training.

Virtually all breeders are willing to share their program innovations and approaches. Awareness about techniques used by other breeders is a valuable aid in program development. Occasionally this information is presented formally at scientific meetings but more often is obtained through casual conversation and program visitation.

The expertise of other scientists is generally employed in achieving program objectives. Organizational skills in constructing a team approach to breeding improvement are valuable.

Figure 10.2. Seeding equipment for experimental sorghum plots. Each row can be seeded to a different selection. Each operator, following a predesigned field plan, places the seed from an envelope in the funnel. It is then distributed uniformly in a given length of row. (Courtesy DeKalb AgResearch, Inc.)

EVALUATION

Periodic program evaluation is necessary in maintaining appropriate objective priorities and assessing progress. The breeder informally evaluates the program every day by the amount of material moving through the pipeline, the number of varieties released, the genetic variability in the program, and the surfacing of new problems and challenges.

Periodically, detailed evaluation is necessary to keep the priorities of major objectives ordered correctly. The same groups providing assistance in establishing the objectives will normally participate in formal program evaluation. The breeder will be required to provide a statement of accomplishments in the form of progress reports or other documentation that will aid in evalu-

Figure 10.3. Small plot grain harvesting equipment. This self-propelled plot combine harvests and packages the grain in one operation. It can also be cleaned easily to reduce mixtures between plots. (Courtesy D. L. Keim, South Dakota State Univ.)

ation. The breeder should be ready to explain achievements and difficulties to growers, administration, and management in order to keep the program operating. Program evaluation can be a highly rewarding experience and will aid significantly in the determination of future action.

Changes in objectives must continually be considered and made if ample justification such as a major pest outbreak is present. However, a breeder will sometimes change objectives too frequently with an accompanying reduction in program efficiency. The problem in changing objectives too often, without a well-planned course of action, is the length of time necessary to produce a new variety or hybrid. Each directional change represents an interruption in the natural flow of material in the breeding pipeline. The breeder should be ready to alter objectives if there is justifiable cause, but alteration for change alone is generally not advisable.

SUMMARY AND COMMENTS

Program design and management offer special challenges to the breeder. If completed properly, maximum productivity and efficiency will be achieved. The implementation and constant evaluation of properly ordered objectives are required in every breeding program. Innovative mechanization can increase results for a given financial expenditure. Often, scientific discipline combinations are necessary to achieve program goals.

The breeders of my acquaintance are, almost without exception, combination scientists and inventors. Because we are in a very specialized business that does not attract mass-produced equipment, we are often forced to make our own. I have yet to meet a breeder who was unwilling to share ideas and innovations with others.

Program design and management is an exciting challenge—it offers the opportunity for interaction with colleagues in the scientific community and producers in the field. I experience great satisfaction in being asked to defend my justification of goals and objectives while remaining flexible enough to keep a dynamic program that can change when necessary. Because there is so little information on breeding program management, it remains a very individualized technique and one of the many fascinating plant breeding areas.

11

BULK BREEDING

Groups of unseparated individuals called bulks can be used in population manipulation. A natural bulk is a wild population or a land race variety that has experienced only minimal selection. In contrast, an artificial bulk is constituted with hybrid progenies of specific parental lines or a composite of selections from among crosses.

Bulk populations can be handled in both self- and cross-fertilized species. Reduced selection pressure intensity in any generation or population accompanied by large numbers of individuals within each population result in bulk concepts being used primarily in agronomic crops with relatively short reproduction times. In general, bulk breeding is very inexpensive since large numbers of individuals can be handled with a small amount of labor. As in all breeding systems, however, bulk methods have advantages and disadvantages. This chapter covers opportunities, problems, and techniques in bulk breeding.

Prior to considering specific systems, we must define population composition terms. Heterozygosity and homozygosity are conditions at particular loci and have been discussed earlier. Heterogeneity and homogeneity will appear in the following discussion. Homogeneity is the condition where a population contains only one genotype, while heterogeneity describes a population of two or more different genotypes. A population is heterozygous and homogeneous if all individuals are heterozygous but exactly alike genotypically. On the other hand, a homozygous heterogenous population would be one with only homozygous individuals but of different genotypes. These terms are important in studying the genetic implications of certain bulk systems.

POPULATION DEVELOPMENT

To practice bulk breeding, a population must be available. Since natural populations are already in existence, no effort is required on the part of the breeder in their production. Finding them involves plant exploration described in Chapter 7. The subject of this section is the artificial creation of bulks by the breeder.

Parent Selection

If the breeder is creating a population through crossing, parental selection obviously plays a key role in the contribution of desired genetic characteristics. As in all breeding efforts, program goals provide valuable input into parental selection. For example, if some form of disease resistance in combination with particular quality properties is desired, then pairs of parents must be identified to contribute variability for one or both characters. Effective parental choice is based on the heritability of the traits and the amount of genetic variability among the parents. If a character is highly heritable, parents with desirable genotypes can easily be identified. On the other hand, low heritability makes the identification of good contributing parents quite difficult. Also, if all parents have the same genotype—that is, no genetic variability—then no progress can be expected. This problem is often overcome by introducing a large number of different parents into the crossing program, even though they may have about the same phenotype. The breeder, in this case, is hoping for transgressive segregation.

Production of Bulks

Bulk population development may be approached in several ways. First, consider self-fertilized crops. Remember that genetic recombination is hard to generate and heterozygosity difficult to maintain in self-fertilized species. Considerable effort must be put forth to generate the desired genetic variability. This can be accomplished by multiple crossing of several parents to produce segregating populations. Crossing all parents in all possible combinations (a diallel cross) may be used. Jensen (5) proposes diallel selective mating (DSM) to produce bulk populations for further selection. In the DSM program he suggests mating as many parents contributing specific characteristics to the population as possible. The F_1 plants are then intermated, resulting in populations heterogeneous and heterozygous for many different genes. The system leads to extensive recombination among and within linkage groups. This is an advantage if undesirable linkages are present, but tends to destroy blocks of genes for general adaptation if widely diverse genotypes are used in the original crosses. The DSM breeding technique is discussed again in Chapter 12.

Since crossing is a time consuming task, breeders have occasionally introduced genes for male sterility (Chapter 16) into the population so natural outcrossing will occur. This has been especially common in barley. At the time of selection the male sterility genes are removed and the crop returns to a self-fertilized condition.

Bulk populations can also be artificially generated in naturally cross-fertilized species. Here heterozygosity and heterogeneity are easy to maintain and, in fact, the former may be necessary for species vigor. To create bulk populations, desirable plants are identified and placed in a crossing environment where they are forced to mate among themselves. Isolation either by geographic distance or mechanical means is required. In many cross-fertilized crops, loss of vigor or inbreeding depression results from genetically close matings. Inbreeding depression appears to be the result of uncovering deleterious recessives that are masked in the heterozygous condition but are expressed in the homozygous state, and from the loss of any potential vigor due to heterozygosity itself. Because of inbreeding depression, population sizes must be large enough to maintain genetic vigor in the bulks. The number of parents is dictated by inbreeding depression levels.

Breeders will often composite or bulk part of the early segregates from many crosses to preserve and maintain genetic variability. The composite cross technique can be extremely valuable as a germplasm bank or reservoir for the breeder's use at any time. The material is easily handled because the breeder will generally maintain only one population seedlot for each cross composite. From time to time genes can be added to the composite for maintenance and recombination.

Following the development of the bulk population, alternative methods of selection can be used. As will be pointed out in following sections, the objective is to isolate desirable genotypes from the bulk population regardless of the selection pressure source. It will become apparent that each selection system has advantages and disadvantages, and the breeder must be aware of these in making the proper management decisions in the breeding program.

NATURAL SELECTION

Several possibilities exist for selection in bulk populations. Keep in mind that every plant in a bulk, unless they are widely spaced, becomes a part of the environment of each of its neighbors. Plant interaction will play a key role in determining the success of bulk breeding systems.

Natural selection has offered an attractive option in the manipulation and utilization of bulk populations in breeding programs. If nature will select the most desirable genotypes, a simple and inexpensive method of plant breeding results. We know, for example, that environmental factors such as temperatures, disease susceptibility, and photoperiod requirements place tremendous selection pressure on populations (Fig. 11.1). Genetic advance will be determined by the reproductive frequency of desirable genotypes.

Considerable experimental work has been done on population dynamics under natural selection. A classical study was conducted in barley by Harlan and Martini (4). They started with an equal mixture of barley varieties and exposed the population to a wide array of environments. After several years of natural selection, the proportion of each variety remaining was determined. Data are presented in Table 11.1. In some environments, the varieties with good agronomic performance survived, and in others they were lost, pointing out that natural selection does not always identify high producing genotypes. Suneson (8) also conducted a population study using natural selection on a mixture of adapted

Figure 11.1. Severe natural selection in winter wheat. The selection nursery was exposed to cold temperatures and wind. The remaining solid rows are very winterhardy winter rye checks. Only a few wheat plants have survived. (Courtesy D. G. Wells, South Dakota State Univ.)

barley varieties. After 16 years of selection one of the poorest yielding varieties in the original mixture made up 88 percent of the population. One of the extremely high yielding varieties had been completely eliminated. This apparent discrepancy between expected and observed results can be explained by the fact that environmental interaction occurs at all stages of plant development in natural selection systems. For example, those plants that were more aggressive in the seedling and early plant development stages might not necessarily be the most productive in a pure stand. Also, years vary greatly and genotypes favored one year may be lost the next.

Natural selection may result in the concentration of both desirable and objectionable genes. Marshall (7) in a natural selection winter survival study with oats showed that significant genetic advance could be made for winterhardiness but increased frequency of undesirable genes for tallness and lateness also occurred. He suggests that controlled selection pressure during one

Table 11.1 Final Census Showing Effect of Natural Selection in a Mixture of Barley Varieties Grown at 10 Locations for 4 to 12 Years, Recorded as the Number of Plants of Each of 11 Varieties Found in a Population of 500 Plants

	Number of Plants of Each Variety in the Year When Last Grown									
Variety	Arlington, Va., 1928	Ithaca, N. Y., 1936	St. Paul, Minn., 1934	Fargo, N. Dak., 1930	North Platte, Nebr., 1932	Moccasin, Mont., 1936	Aberdeen, Idaho, 1936	Pullman, Wash., 1930	Moro, Oreg., 1934	Davis, Calif., 1928
Coast and Trebi	446	57	83	156	224	87	210	150	6	362
Gatami	13	9	15	20	7	58	10	1	0	1
Smooth Awn	6	52	14	23	12	25	0	5	1	0
Lion	11	3	27	14	13	37	2	3	0	8
Meloy	4	0	0	0	7	4	8	6	0	27
White Smyrna	4	0	4	17	194	241	157	276	489	65
Hannchen	4	34	305	152	13	19	90	30	4	34
Svanhals	11	2	50	80	26	8	18	23	0	2
Deficiens	0	0	0	1	3	0	2	5	0	1
Manchuria	1	343	2	37	1	21	3	1	0	0

Source: Harlan and Martini (4).

or more early generations may be necessary to concentrate desirable alleles for characters other than those under natural selection. Natural selection can be an effective force in varietal development. For example, Grimm, one of the most winterhardy alfala varieties ever produced, was the result of natural selection for cold tolerance in a northern latitude. Natural selection pressure, then, is a series of subtle or dramatic environmental interactions with plant genotypes. The effectiveness of the selection system is not well understood and seems to produce conflicting results depending on the experimental materials and environment.

ARTIFICIAL SELECTION

The greatest value of bulk populations is in their supply of genetic diversity for selection by the breeder. Natural populations are a logical source of genetic variability for the evolution of crop species and development of varieties. Historically, people have utilized naturally occurring populations to pick satisfactory plant types for use and advance to the next generation. Mass selection is the identification of superior plants and their subsequent bulking to form the next generation. In natural populations the identification of superior individuals is based entirely on phenotypic expression.

While mass selection in natural populations has been an effective evolutionary tool, two problems must be considered. In self-fertilized species there is a limit to the genetic diversity that can occur in a natural population selfed without selection over a long period of time where highly homozygeous heterogeneous populations are developed. In a classic study of selection potential in self-fertilized species, Johannsen (6), using beans and starting with local seed sources, demonstrated that seed size could be changed only as long as genetic diversity existed in the population. When he obtained homozygous plants for maximum and minimum seed size, selection was no longer effective. This work, resulting in the pure-line theory, is a valuable principle to consider when applying selection pressure in self-pollinated crops.

A different problem arises with mass selection in cross-fertilized crops. As mentioned earlier, many cross-fertilized crops experience inbreeding depression when homozygosity is promoted. The selection of a very few genotypes for advancement to the next generation would narrow the range of genetic diversity in a cross-fertilized crop very quickly. Therefore, the success of mass selection in species that experience severe inbreeding depression would depend on maintaining a large number of individuals from generation to generation so that high levels of heterozygosity would be maintained.

Selection progress may be much less limited in cross-fertilized than in self-fertilized plants. For example, Dudley et al. (1) reports a 70 year selection study in maize for oil and protein concentration in the kernel. Starting with an open pollinated variety, mass selection with some modifications to reduce inbreeding depression was practiced. Selection for high oil (Illinois High Oil or IHO), low oil (ILO), high protein (IHP), and low protein (ILP) continued to produce progress even after 70 generations. In addition, reversing the selection pressure in each line (RHO, RLO, RHP, and RLP) resulted in opposite direction progress. The experiment is summarized in Figures 11.2 and 11.3. Apparently, much more variation for these characters was present in the germplasm than had originally been anticipated. An unfortunate relationship exists in this material, however, as yield reduction occurred with improved oil and protein. The experiment is being continued as a valuable long-term selection study.

Mass selection can be effective in genetic improvement of crops. Gardener (2) demonstrated yield improvement in corn for about 16 generations until the genetic potential for improvement had apparently been exploited. Hansen et al. (3) showed that after several cycles of mass selection in alfalfa, resistance to rust, common leaf spot, leaf hopper, and spotted alfalfa aphid was increased. Mass selection has also proven valuable in sugar beets where sugar content has been increased from 11 percent in early genotypes to over 18 percent in modern varieties. In modern plant breeding systems mass selection is occasionally used to purify varieties that have been contaminated by off types. This is accomplished by roguing out the undesirable plants and bulking the remainder.

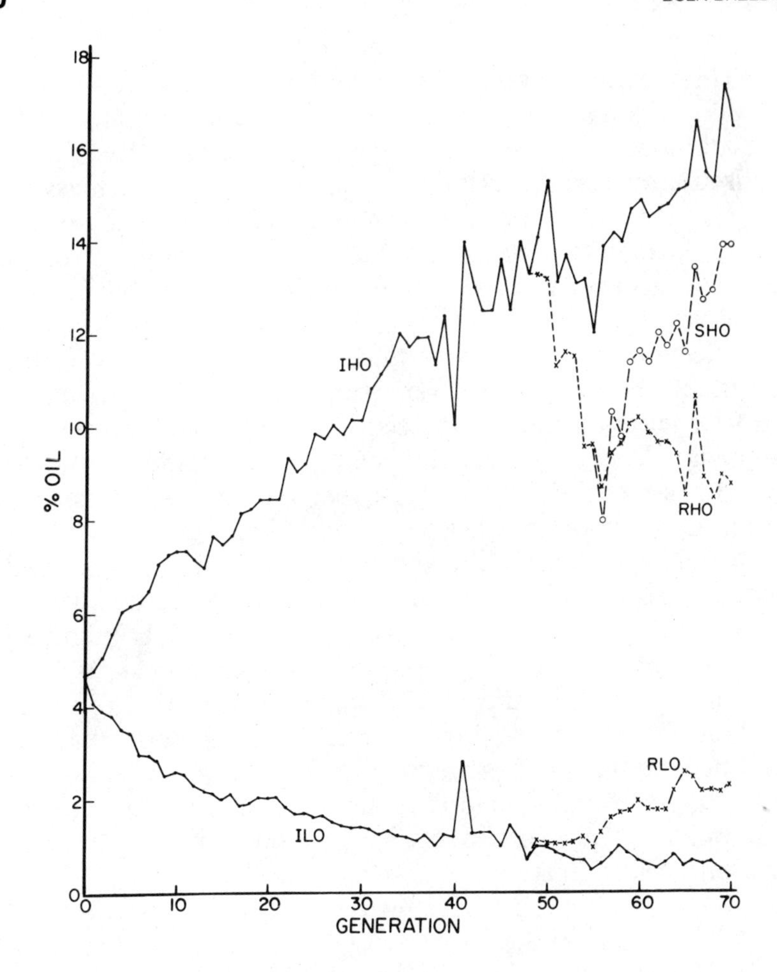

Figure 11.2. Seventy generations of selection progress for oil content in maize. IHO—Illinois High Oil, ILO—Illinois Low Oil, RHO—Reverse High Oil, RLO—Reverse Low Oil, SHO—Switchback High Oil. (From Seventy generations of selection for oil and protein in maize, 1974, p. 188. By permission of the Crop Science Society of America.)

Mass selection can be accomplished for one trait by selecting for another. This is sometimes called secondary selection. Tipyawalee and Frey (9) used seed size in oats as a selection tool to improve crown rust resistance. The success of the selection system was based on the concept that resistant plants will produce better seed in terms of size and development than susceptible plants will.

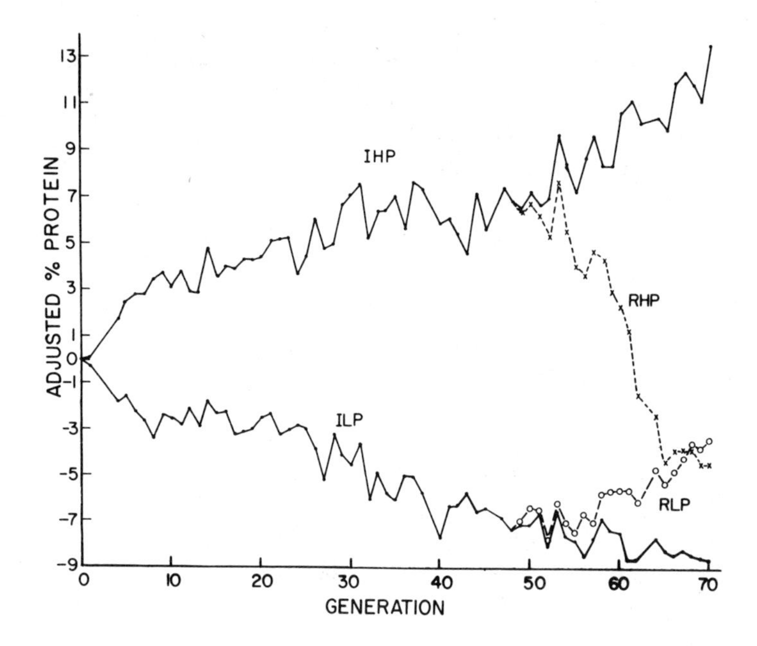

Figure 11.3. Seventy generations of selection progress for protein content in maize. IHP—Illinois High Protein, ILP—Illinois Low Protein, RHP—Reverse High Protein, RLP—Reverse Low Protein. (From Seventy generations of selection for oil and protein in maize, 1974, p. 189. By permission of the Crop Science Society of America.)

By exposing populations to artificial crown rust epiphytotics and then screening the resulting seed, they were able to significantly increase rust resistance in the population. Breeders continually evaluate potential mechanical selection systems that can be superimposed to identify the most desirable genotypes in bulk populations. Reduced height, for example, can be selected by clipping the tall plants just prior to seed production.

Time of selection becomes important when selecting in cross-pollinated crops. If unwanted plants can be removed prior to flowering, the resulting genotypes will be a combination of male and female gametes from desired plants. If selection cannot be accomplished until after flowering, concentration of desirable genes will be reduced since there is no control over the male gametes.

The success of selection systems depends on the ability of the breeder to judge the correct phenotype, and the heritability of the traits under consideration. In the 70 generation corn selection experiment, phenotype identification was quite precise since it involved mechanical measurements for oil and protein. Selection progress dependent on visual phenotype evaluation is subject to more error, especially if heritability is low. Genotypic variation for characters such as leaf number, vigor, and yield components is difficult to estimate visually. With low heritability characters, gains are slow if they can be achieved at all. Selection is particularly effective on highly heritable traits that can be easily identified in the plant population.

SUMMARY AND COMMENTS

Bulk methods are the oldest plant breeding systems in existence. Natural and artificial selection techniques can be extremely effective in improving the genotype composition of populations. Bulk population programs are generally inexpensive and easy to operate, but inbreeding depression can be a serious problem in bulk breeding cross-fertilized crops. The identification of pure lines is the end of genetic progress in bulk handling of selfed crops. Natural selection has the advantages of being very inexpensive and containing the potential to identify genotypes that can withstand environmental stress. It does not insure that the highest yielding genotypes will be selected. Artificial selection can be done rapidly and inexpensively, but again the problem of correct genotype identification becomes critical.

Winter wheat is subjected to environmental stresses of cold and drought. I have used natural selection in bulk population breeding systems with varying degrees of success. Froid, one of the most hardy winter wheats developed in Montana, was identified by natural selection in the eastern part of the state. Drought, on the other hand, seems to present a much more difficult problem. We experience such a variation in environmental conditions from year to year that natural selection techniques seem to work very poorly. A seed screening system has been imposed on populations that have been produced under drought stress with the idea that better plants produce better seeds. The results of this selection system still remain to be evaluated. I am confident, however, that some mechanism will be found to better identify those genotypes that can perform well under difficult conditions. It is entirely possible that some form of bulk population manipulation will play a key role in screening large numbers of genotypes.

REFERENCES

1. Dudley, J. W., R. J. Lambert, and D. E. Alexander. 1974. Seventy generations of selection for oil and protein concentration in the maize kernel. pp. 181-212. In J. W. Dudley (ed.). *Seventy generations of selection for oil and protein in maize.* Crop. Sci. Soc., Madison, Wisc.

2. Gardner, C. O. 1978. Population improvement in maize. pp. 207-228. In D. B. Walden (ed.). *Maize breeding and genetics.* Wiley, New York.

3. Hansen, C. H., T. H. Busbice, R. R. Hill, Jr., O. J. Hunt, and A. J. Oakes. 1972. Directed mass selection for developing multiple pest resistance and conserving germplasm in alfalfa. *J. Environ. Qual.* 1:105-111.

4. Harlan, H. V., and M. L. Martini. 1938. The effects of natural selection in a mixture of barley varieties. *J. Agric. Res.* 57:189-199.

5. Jensen, N. F. 1970. A diallel selective mating system for cereal breeding. *Crop Sci.* 10:629-635.

6. Johannsen, W. 1903. Veber Erblichkeit in Populationen und in Reinen Leinen. Jena: Gustav Fisher. Translation of summary in *Classic Papers in Genetics,* Prentice-Hall, Inc., Englewood, N.J., James A. Peters, Ed., 1962.

7. Marshall, H. G. 1976. Genetic changes in oat bulk populations under winter survival stress. *Crop Sci.* 16:9-15.

8. Suneson, C. A. 1949. Survival of four barley varieties in a mixture. *Agron. J.* 41:459-461.

9. Tipyawalle, D., and K. J. Frey. 1970. Mass selection for crown rust resistance in an oat population. *Iowa State J. Sci.* 45:217-231.

12

PEDIGREE BREEDING

Pedigree breeding, in which ancestral lineage is recorded, has been used since the rediscovery of Mendel's laws. In pedigree breeding programs care is taken to keep accurate records so that each selection can be traced back to the original parental combination. The pedigree system also allows comparisons among relatives at each generation level. This is a pedigree system in its strictest sense, and modifications and adjustments have evolved over the years. Pedigree breeding methods are most often applied to annual self-fertilized seed crops, but can be used in other species as well.

When using the pedigree approach, development of the cross and pedigree language becomes important. Prior to the utilization of computer systems, parental information was written using algebraic symbolism as follows: [(Parent A × Parent B) × Parent C] × Parent D. This is a complex cross, which happens frequently, and was made chronologically with A × B first, the F_1 or some descendant from A × B crossed with C, and D entering the pedigree last. Repeated use of a parent in a backcrossing system was written A × B^3 indicating three doses of parent B. Following the introduction of computerization, symbol changes to computer language were necessary. A new system suggested by Lamcraft and Finlay (13) is illustrated as follows: ((A*B)*C)*D or in the case of the backcross A*(B)3. All information can now be handled by the computer. To reconstruct the parental combination sequence, the most recent parent will appear on the right. Pedigrees can thus be stored and recalled electronically.

Selection numbers identify the source of the individual selected in each generation. For example, the pedigree for two different selections from the cross A*B might be (A*B)/1/7/3/5 and (A*B)/1/7/9/15. The "/" signals that the following number is a selection identification. In this example, the "1" indicates plant number 1 in the F_1, the "7" indicates plant 7 in the F_2 generation, the "3" or "9" indicates the specific plants or rows in the F_3 generation, and "5" or "15" indicates the F_4 source. These two selections both originated from the same F_1 and F_2 plants, but then diverged at the F_3 generation. Many breeding programs develop computer systems to automatically advance the pedigree for each round of material selected. In other cases this is still done by hand. Extensive bookkeeping detail is associated with pedigree breeding.

PARENT SELECTION

The parents define the genetic variability available to the program and their selection is a decision of critical importance. If the breeder chooses the wrong parents, the probability of achieving genetic advance is reduced. If, on the other hand, the breeder chooses a very small well-defined set of parents with a very narrow germplasm base, the chances of finding desirable genetic segregates for many characters may be quite low. Transgressive segregation between two mediocre parents, while not particularly easy to predict, may be a valuable possibility. Highly heritable characters are much easier to deal with than those with low heritability. Computer programs to identify the correct parents to use in crosses for specific objectives have met with only limited success. Each breeder always has the responsibility to study all potential parent material, and make choices based on past experience and knowledge regarding heritability of traits important to specific breeding goals. A breeder should not be particularly discouraged with some poor parental choices, especially if a program is just being initiated. With our current level of knowledge, the accurate prediction of every cross combination outcome is impossible. On the other hand, wise parent selection will help improve the odds for favorable results.

Breeders are constantly faced with the decision of how many parents to use and number of crosses to make. Some breeders may make as few as four or five crosses annually and generate extremely large populations with the intent of detailed study and intensive selection within each population. Others, by contrast, may make several hundred combinations annually, maintain small numbers within each population, and discard whole crosses early in the selection program. There are advantages and problems with both approaches. For the breeder who has made few crosses, the problem of accurately identifying correct parental combinations is very critical since wrong parental choices leave the program with few populations containing desirable segregates. However, extensive evaluation of many progeny in a few crosses can lead to identification of low frequency desirable genotypes. If parents are selected for quite similar phenotypes and genotypes, the likelihood of transgressive segregation decreases. These are the tradeoffs in a conservative program.

Other breeders may make many crosses knowing that the population numbers within each cross will necessarily be small because of financial constraints on overall program size. The total number of progeny in the breeding program will be similar, only the distribution by parents will be different. A wide array of genetic diversity exists in this type of program. Potential transgressive segregation can be expected quite often since a great deal of genetic diversity, observed or unobserved, may exist in the parental array. There is a problem, however, because large numbers of crosses reduce the number of segregates that can be handled within each cross and consequently reduce the probability of finding low frequency highly desirable phenotypes within a cross. Thus, the breeder who makes many crosses is also faced with a series of tradeoffs that must be considered in program design. Most breeders will accept a compromise between the two extremes and try to capitalize on the advantages of both while minimizing the disadvantages. Techniques developed to do this are discussed in the following sections.

HYBRIDIZATION

Several alternatives exist for combining the genetic variability from selected parents. A series of single crosses may be made between parents, resulting in the number of populations equivalent to the parental combinations. Some parents may be repeated in several crosses. This may be satisfactory for the goals of the program but it may also be desirable to incorporate additional diversity into each population by making three-way and multiple crosses.

In Chapter 11, the DSM system was mentioned. Tremendous amounts of genetic diversity are introduced using this system or some variation of it. Jensen (10) suggests grouping parents so that highly diverse populations for specific goals can be produced. For instance, high protein may be a general goal of one population and many parents with apparent high protein will be combined in the mating system. Another goal may be short plant stature and a third some form of plant adaptation such as winterhardiness. His proposal probably represents the maximum infusion of genetic diversity in pedigree breeding and some breeders will argue that the diversity is too great to be handled by practical selection techniques.

Highly diverse populations with specific goal orientation can also be used as parent building sources for future crossing programs. In fact, all hybridization systems are normally a source of future parents since plants improved for several characters, but still deficient in some, are usually recycled through the breeding system. Source populations are sometimes intentionally created and maintained by bringing together blocks and concentrations of genes for specific characters.

Before cross combinations are made, the numbers of populations that should be entered into the breeding pipeline each year must be determined. Limits of efficient selection dictate total progeny numbers since genetic diversity is of no value unless it can be exploited. Therefore, crossing programs must stay within the economic framework of selection capacity, which in turn is dependent on the selection methods, environmental variability, and the heritability of important traits.

Most breeders, through experience, establish program limits in terms of numbers of plots, numbers of locations, or other physical criteria that define capacity to handle material. Each breeder must operate within a given framework of financial resources, and inadequate planning results in the inability to exploit genetic variability.

Prior to the selection technique discussion, terminology used throughout the remainder of the chapter must be developed. A head row and plant row are the results of a head or seed subsample respectively obtained from a single plant. For example, F_3 head rows would be individual rows propagated by selecting heads from F_2 plants. F_3 plant rows would be approximately equivalent but would have been generated by a random sample of seed from the entire plant. A family refers to a group of progeny that have all originated from a previous single source. For example, an F_4 family of 20 rows would have originated either by head or plant selection from a single F_3 progeny.

POPULATION SIZE

In any breeding program the maximum level of genetic diversity occurs in the F_2 population. Theoretically, an F_2 population of adequate size will contain every genotypic combination in its correct frequency for all characters segregating between the parents. The breeder must determine population size for each F_2 in order to insure a reasonable probability that the desired genotypes will occur. The population size depends on the complexity of the genetic system. Stevens (17) states, for example, that if 21 genes are segregating in a tomato population, it would take over 420,000 acres of F_2 plants to be assured that all the individual genotypes occur at least once. The objective becomes one of producing adequate population sizes to provide the breeder with a reasonable (but not absolute) probability of finding desirable genotypes. Here the breeder pits skill and experience against the odds generated by the genetic control of the system and its interaction with the environment. In general, as heritability goes down, the odds for a favorable outcome are reduced because of less relationship between the phenotype and genotype. The real problem is to recognize the correct genotype when it occurs. Thus the relationship between genotype and phenotype becomes all important in a breeding program.

Mather (14) has generated probabilities, based on statistical theory, of a genotype occuring in a particular population (Table 12.1). With a single locus, suppose we wish to know how many F_2 individuals we should produce to attain a 99 percent probability of having at least one homozygous recessive genotype. Since the recessive genotype makes up one-quarter of the population, we would need to grow 16 individuals to have a 99 percent probability of having at least one. If we are willing to settle for a 95 percent probability, then only about 11 individuals are necessary. Note that as the desired genotype fraction of the population declines, the required number of individuals increases drastically. In the three gene segregating system, a homozygous recessive representing 1/64 of the population would require 292 individuals for a 99 percent occurrence probability, but the necessary number for a 95 percent probability is only 190. The chance the breeder is willing to take has a very significant role in population size determinations. More recent work agreeing basically with these calculations

Table 12.1 Population Size Necessary to Have Given Probabilities of Individuals From the Designated Fraction Present.

	Level of Probability	
Fraction Expected	0.95	0.99
½	4.3	6.6
⅓	7.4	11.4
¼	10.4	16.0
⅛	22.4	34.5
1/9	25.4	39.1
1/16	46.4	71.4
1/27	79.4	122.0
1/32	94.4	145.1
1/64	190.2	292.4

Source: Reprinted by permission from *The Measurement of Linkage in Heredity,* K. Mather. Wiley, 1951.

has been reported by Sedcole (16) in which he provides several alternative methods for calculating specific population fraction occurrence probabilities.

A compromise must be made between the number of cross combinations and the population size within each cross. The breeder must continually reduce numbers through selection if genetic gain is to be expected in each generation. Accurate selection, the identification of desirable genotypes, serves as the key to a successful program. The more effective the selection program, the more likely the success in producing desirable genotypes. Because of selection impact on a program, much study has been devoted to the development of effective and innovative selection techniques.

SELECTION

Selection can be handled in one of two general ways. First, classical pedigree selection without yield testing can be conducted. The selections can be either random or based on phenotypic evaluations primarily through visual selection.

Visual selection is the identification of valuable genotypes based on mental images of desirable plant types (ideotypes). Components of an ideotype may include such things as plant height, leaf size and shape, tillering capacity, vigor, color, and pest reaction. Different breeders will use different ideotypes based on their experience and available scientific information. When using ideotypes in selection, populations are biased toward the breeder's preconceived idea of a perfect plant. Serious problems result if the conceptions are wrong since incorrect genotypes are selected and propagated. Visual selection usually starts in the F_2 population on an individual plant basis and can be continued through later generations of head and plant rows. It is particularly effective in rapidly reducing population size because of the speed with which it can be carried out. Strict pedigree population management lends itself to small numbers of crosses with heavy sampling in each population. Utilizing head row or plant row selection prohibits large volume testing for such items as quality, and screening may be limited to microtests.

The other extreme of population management involves the use of bulk testing, a variation of bulk population handling described in Chapter 11. Considerable attention has been given to the possibility of yield testing F_2 population bulks to evaluate average performances and provide an identification of genetic potential within each population. If accomplished successfully, the breeder could then generate many more segregating populations than could be used, but would be able to identify and select only the most productive ones. Harlan et al. (8) suggested the possibility of using F_2 bulk populations in performance predictions for further selection in barley and indicated that this technique could indeed be used to identify highly productive populations. Other data, however, by Kaltan (11) with soybeans and Atkins and Murphy (1) using oats suggested that the yield relationship of later generation performance to early generation bulk testing was not strong. Cregan and Busch (3) in 1977 reported that bulk early generation yield tests in spring wheat could identify superior crosses even when parents were selected for high yield potential.

Bulk testing has also been considered for generation advancement beyond the F_2 where populations and individuals to be advanced from generation to generation are identified entirely on some measurement of performance such as yield. The breeder pays little heed to morphological traits of the plants but instead bases selection on data collection and interpretation. The theory is that performance values will identify those populations or families having the highest probabilities of containing more productive genotypes.

Much experimental evidence has been obtained regarding the relative merits and success levels of strict pedigree and modified bulk selection. The pertinent question is the comparative efficiency of visual selection and performance testing, compared with each other and with random chance. Studies have been carried out in wheat by McGinnis and Shebeski (15), Knott (12), and Townley-Smith et al. (18), in oats by Frey (4), in barley by McKenzie and Lambert (20), and in soybeans by Voight and Weber (19) and Hanson et al. (7). In each study the plants or lines were visually rated and then measured for yield. Unfortunately, the results are not clear in identifying the superior selection technique. Several studies showed that visual selection was effective in identifying those individuals that had high performance capacity while in others visual selection was no more effective than randomly picking the lines. Apparently difficulties arise when the genotype-environment interactions change. High performance genotypes can be identified visually in some environments but not in others.

As with visual selection, there appears to be no clear-cut answer with respect to the validity of bulk yield testing for the ultimate selection of superior genotypes. If heritability is low and there is a very high degree of environmental interaction, then the value of the test is likely to be low since the environment can vary extensively from year to year and possibly result in erroneous selection decisions based on an abnormal situation. Another weakness lies in the fact that tests are based on the mean value and do not provide the range and distribution pattern within each population. In populations discarded on their mean values, low frequency high yielding individuals may be lost because a high proportion of poor individuals pulls the mean down. On the other hand, if each population has approximately the same distribution pattern, then the test should be valid in identifying crosses with good probabilities for containing high performance selections. A third weakness lies in the fact that some superior genotypes will only express their potential in pure stands but may be suppressed in genotypic mixtures. The question regarding the validity of bulk testing remains unanswered, but it will undoubtedly receive considerable investigation in the future and is currently being used in a number of small grain programs in connection with modified pedigree advancement.

Many breeders compromise by bulk testing one or two generations followed by visual or pedigree selection for several cycles. The theory here is to combine the selection pressure potential of both systems, but not entirely rely on one or the other. Another variation includes the visual selection of highly heritable characters such as disease resistance that can be observed readily, and bulking the selected individuals for subsequent performance testing. This results in biasing the population toward observable desirable characteristics but also includes performance evaluations.

This combination has been used extensively and appears to have a good level of efficiency.

Another modification in population management is a system called single seed descent, first described by Goulden (5) in 1941, and later discussed by Brim (2) in soybeans. It is a quick technique to generate homozygous populations in self-fertilized species. The method simply involves advancing one seed from each F_2 plant in a population to the F_3 generation that then is similarly advanced to the F_4 generation, and so on. By using techniques such as harvesting soon after embryo formation, several generations per year can be produced. For example, Grafius (6) describes a system of barley culture in which 8000 plants could be grown in a 1.1×8 m bed using seed harvested two weeks after anthesis. With this production system a large number of progeny can be produced each season in a greenhouse. Using single seed descent, it is possible to very rapidly produce populations of homozygous lines that represent a wide sampling of the F_2 variability. After reaching the desired level of homozygosity, the plants can be increased individually and any one of several selection systems applied. The net result is to test homozygous lines representing most of the genetic variability available in the population soon after the cross is made. This system is becoming widely accepted and being used extensively by many breeders of annual self-fertilized crops. In using single seed descent, care must be taken to sample enough plants so variability is not lost, since the final number to be tested is no more than the number of F_2 plants sampled.

A final variation involves the application of environmental stress for such characters as drought or disease and subjecting the resulting populations to mechanical mass selection by seed screening. The theory suggests that natural selection should first occur and weed out those genotypes that might be poorly adapted, leaving the population with individuals having good adaptation and yield potential. Following one or two generations of natural selection, the populations can be subjected to conventional pedigree techniques. Evaluation is then conducted for yield and other program objectives to further identify desirable segregates. This program has the advantage of selecting for those characters that are extremely quantitative but, again, environmental variation can lead to erroneous selection decisions.

Pedigree breeding can be used as a means of producing varieties directly for the commercial market. It can also be used in the development of inbreds for use in hybrids (Chapter 15), for the development of synthetic varieties (Chapter 14) or as the base breeding program for multiline varieties (Chapter 13). The student is encouraged to remember that many breeding programs can be used as the building blocks for other breeding techniques that will result in varieties or hybrids. There are no strict limitations in the way any single technique is handled or used.

SUMMARY AND COMMENTS

The pedigree system of plant breeding offers a wide array of challenges to the breeder. As in all breeding techniques, correct parental combinations must be made to produce desirable segregates. Parental decisions should always consider transgressive segregation possibilities.

A number of options and modifications ranging from bulk testing to strict pedigree and visual advancement are open to the breeder in population management and selection. The breeder must decide the correct system for a program based on experience and knowledge of the characters. Modification such as single seed descent offer the possibility of innovative selection techniques.

I have participated in many spirited debates with my colleagues regarding the relative merits of visual selection and bulk testing. Personally, my confidence in accurately identifying highly productive genotypes by visual selection for stress environments is quite low. I do much better for very productive environments, however. Modernization of field plot seeding and harvesting equipment has opened up new possibilities in performance testing large numbers, and many of us are now revising and updating selection systems. The new innovations add an exciting dimension to the plant breeding game and it will be interesting to see what our programs are like at the end of the century.

REFERENCES

1. Atkins, R. E., and H. C. Murphey. 1969. Evaluation of yield potentialities of oat crosses from bulk hybrid tests. *Agron. J.* 41:41–45.

2. Brim, C. A. 1966. A modified pedigree method of selection in soybeans. *Crop Sci.* 6:220.

3. Cregan, P. B., and R. H. Busch. 1977. Early generation bulk hybrid testing of adapted hard red spring wheat crosses. *Crop Sci.* 17:887-891.

4. Frey, K. J. 1962. Effectiveness of visual selection upon yield in oat crosses. *Crop Sci.* 2:102-105.

5. Goulden, C. H. 1941. *Problems in plant selections.* Proc. 7th Int. Genetical Congress, Edinburgh. pp. 132-133.

6. Grafius, J. E. 1965. Short cuts in plant breeding. *Crop Sci.* 5:377.

7. Hanson, W. D., R. C. Leffel, and H. W. Johnson. 1962. Visual discrimination for yield among soybean phenotypes. *Crop Sci.* 2:93-96.

8. Harlan, H. V., M. L. Martini, and H. Stevens. 1940. *A study of methods in barley breeding.* U. S. Dept. Agric. Tech. Bull. No. 720.

9. Hinz, P. N., R. Shorter, P. A. DuBose, and S. S. Yang. 1977. Probabilities of selecting genotypes when testing at several locations. *Crop Sci.* 17:325-326.

10. Jensen, N. F. 1970. A diallel selective mating system for cereal breeding. *Crop Sci.* 10:629-635.

11. Kalton, R. R. 1948. *Breeding behavior at successive generations following hybridization in soybeans.* Iowa Agric. Exp. Sta. Res. Bull. No. 358.

12. Knott, D. R. 1972. Effects of selection for F_2 plant yield on subsequent generations in wheat. *Can. J. Plant Sci.* 52:721-726.

13. Lamcraft, R. R., and K. W. Finlay. 1973. A method for illustrating pedigrees of small grain varieties for computer processing. *Euphytica* 21:56-60.

14. Mather, K. 1951. *The measurement of linkage in heredity.* Wiley, N. Y.

15. McGinnis, R. C., and L. H. Shebeski. 1973. *The reliability of single plant selection for yield in F_2.* Proc. 4th Int. Wheat Gen. Symp. pp. 410-415.

16. Sedcole, J. R. 1977. Number of plants necessary to recover a trait. *Crop Sci.* 17:667–668.

17. Stevens, M. A. 1973. The influence of multiple quality requirements on the plant breeder. *HortScience* 8:110–112.

18. Townley-Smith, T. F., E. A. Hurd, and D. S. McBean. 1973. *Techniques of selection for yield in wheat.* Proc. 4th Int. Wheat Gen. Symp. pp. 605–609.

19. Voight, R. L., and C. R. Weber. 1960. Effectiveness of selection methods for yield in soybean crosses. *Agron. J.* 52:527–530.

20. McKenzie, R. I. H., and J. W. Lambert. 1961. A comparison of F_3 lines and their related F_6 lines in two barley crosses. *Crop Sci.* 1:246–249.

13

BACKCROSS BREEDING

Backcrossing, originally described by Harlan and Pope (8) in 1922, is commonly used in many crops. Backcrossing is defined as the repeated crossing to one parent and is illustrated in Figure 13.1. The repeating parent is called recurrent, the other nonrecurrent. Generations are designated with the first cross as the F_1 and each backcross assigned BC_1, BC_2, and so on. Prior to computerization, the pedigree for the third backcross would have been written $N \times R^4$. With computer processing this pedigree would now be N*(R)4. Following completion of the backcrossing, generation and selection identification continues as a pedigree system. For example, N*(R)4/1/7 identifies the F_1 and F_2 individuals after the third backcross. Note that the backcross generation number is always one less than the number of crosses to the recurrent parent.

The backcross system was originally designed to add highly heritable desirable alleles from a nonrecurrent parent to the genetic background of a recurrent parent. Both the specific allele being added and the genetic background condition must be considered in population composition.

POPULATION DYNAMICS

The genetic background can be considered in terms of the total information percent contributed by either parent. In the F_1 the genetic information from each parent is 50 percent. As we begin to backcross, contribution from the nonrecurrent parent is gradually decreased and the recurrent parent is increased. Population bias through backcrossing is demonstrated by Whitaker (18)

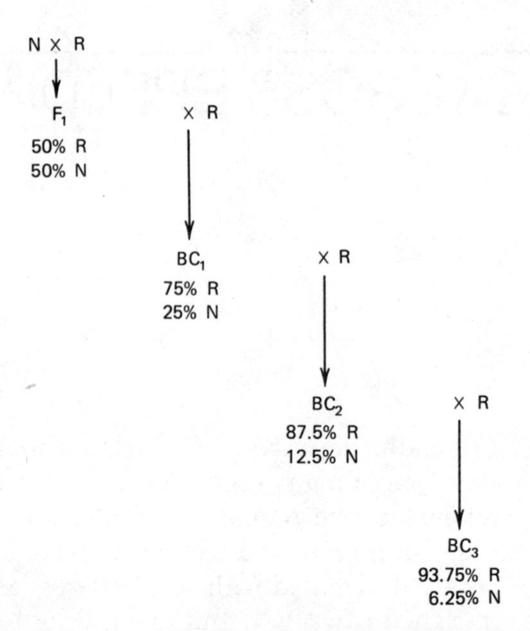

Figure 13.1. Backcrossing. R—recurrent parent, N—nonrecurrent parent. Percentages in progeny indicate average genetic composition.

with gourds in Figure 13.2. The index value, based on the integration of measurements for seven characters, was markedly moved toward the recurrent parent after only one backcross in either direction. The amount of remaining genetic information, *on the average,* from the nonrecurrent parent is reduced by 50 percent with each backcross and is calculated by the formula $(\frac{1}{2})^n$ where n = the number of crosses to the recurrent parent. For example, in the BC_2 the nonrecurrent parent contribution is 1/8 or 12.5 percent.

In the F_1 all loci differing in alleles between the parents will be heterozygous. With backcrossing, increased homozygosity for recurrent parent alleles will occur. The rate with which this takes place can be calculated by the formula $(2^m - 1)/2^m$ where m is the number of backcrosses. For example, in the third backcross 7/8 or 87.5 percent of the segregating loci will be homozygous.

Another population consideration is the proportion of genotypes homozygous for all of the original heterozygous loci. This

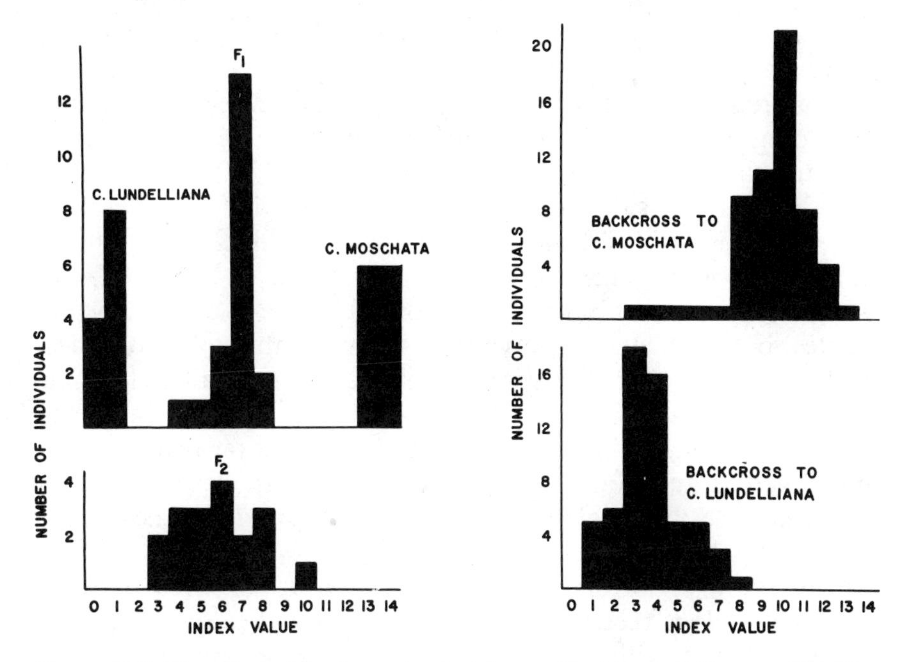

Figure 13.2. The effect of backcrossing in interspecific crosses with *curcubita*. The index value is an accumulated measurement of several plant characteristics. A single backcross to either parent resulted in a pronounced bias in population distribution toward the recurrent parent.

can be obtained by the formula $[(2^m - 1)/2^m]^n$ where n equals the number of heterozygous loci in the original F_1 and m equals the number of backcrosses. With two loci and three backcrosses, the homozygous proportion of the population will be 49/64 or approximately 77 percent. The fact that the recovery rate of the recurrent parent gradually reduces each generation becomes important when the breeder is determining the number of backcross generations necessary to satisfy breeding objectives. In the case where parents are quite similar genetically, a very minimal theoretical amount of recurrent parent recovery may be adequate, while with very diverse genotypes it may take several backcrosses to remove undesirable genetic information. Linkage is a complicating factor. If the desired allele from the nonrecurrent parent is linked with other unwanted alleles, reduced speed of recurrent parent genotype recovery will occur. The removal of the nonrecurrent genotype around the contributed locus is dependent on recombination in that particular chromosome region. The recombination frequency is determined by map distance and other factors including centromere location and chromosome structural abnormalities that reduce crossover events.

We have considered the genetic background of each parent in the composition of the backcross progeny. Of equal importance is the specific management of the allele or alleles to be contributed by the nonrecurrent parent. Since the objective is to add specific alleles to a particular genotype, we must be able to follow these alleles in the crossing program. If the contributed allele is dominant, there is very little problem since its presence can be identified phenotypically in each generation and backcrossing can be continued in an uninterrupted pattern. After the desired number of backcrosses the final progeny can be selfed and the allele stabilized in the homozygous condition.

A slightly more complicated situation arises when the contributed allele is recessive. The F_1 will automatically contain the recessive allele, although it cannot be observed phenotypically. However, in the BC_1 generation 50 percent of the progeny will be homozygous dominant at that locus and 50 percent will be heterozygous. At this point it would be very easy to lose the desired allele and continue backcrossing with no chance of achieving desired objectives. A selfing operation must be conducted to identify homozygous recessive BC_1F_2 progeny. These can then be backcrossed to the recurrent parent and the entire procedure repeated. Note that the basic assumptions of parental genetic background contribution are not being altered. If adequate numbers of crosses in each backcross generation are conducted, the selfing interruptions may be reduced by improved odds of having the recessive allele. The recurrent parent (*AA*) crossed at random to 20 individuals in a population of 1*AA*:1*Aa* will have a much higher likelihood of producing some *Aa* progeny than would five crosses. However, periodic selfing is imperative to be sure that the recessive allele is present. Breeders may occasionally conduct simultaneous backcrossing and selfed progeny testing. Here the backcross program is continued but crossed progeny are also selfed and populations from homozygous dominant genotypes are discarded.

BREEDING FOR SPECIFIC ECONOMIC OBJECTIVES

Backcrossing has been most used for transferring desirable alleles of qualitative traits. One of the most common is disease resistance. The California backcross breeding program in wheat, described by Briggs (3,4) and Briggs and Allard (5) is a classic example. A number of varieties have been generated using Baart as the recurrent parent and incorporating resistance alleles for several diseases. Using highly desirable recurrent parents for backcrossing is a valuable breeding strategy. Baart had extremely good performance in California for most traits except disease resistance, and selection of acceptable backcross genotypes presented minimum difficulties.

The nonrecurrent parent genotype, with the exception of the contributing alleles, can be quite poor since it is eliminated rather rapidly. However, with poor genotypes more backcrosses may be required to remove the unwanted alleles. Four or five backcrosses are commonly used in variety development. Backcross progeny can be selfed during the program and selection practiced for desirable genetic background traits. This will lead to a faster recovery of the recurrent parent as intense selection has the approximate equivalence of one or two additional backcrosses.

Backcrossing has been used in plant species other than cereals for the recovery of desirable traits. In apples, for example, Houg et al. (11) described the use of backcrossing to incorporate apple scab resistance from *Malus floribunda.* Keep and Knight (13) utilized the black raspberry and other *Rubus* species in backcrossing programs for incorporation into the commercial red raspberry of such characters as alternate methods of vegetative propagation, monogenic resistance to aphid virus, high flower numbers per lateral, and stout upright canes. Sprague (15) describes the use of backcrossing in the genotype improvement of inbred lines for use in corn hybrids. Thomas (17) provides extensive detail on the use of backcrossing in several noncereal crops including cotton, tobacco, tomato, and potato.

Each breeding program that utilizes backcrosses must make adjustments to accommodate particular genetic systems and population structures. Quantitative, as well as qualitative, traits can be manipulated through backcrossing. If quantitative traits are desired from a nonrecurrent parent, possibly only one or two backcrosses may be desirable to maintain a large number of contributed alleles. Too many backcrosses would quickly remove a number of the desired genes.

Backcrossing can be used effectively in cross-pollinated as well as in self-pollinated species. If a species experiences severe inbreeding depression with close mating, then backcrossing will have to be conducted on a very limited scale and will likely require the use of many recurrent parent plants to maintain natural vigor. In alfalfa, backcrossing has been used to improve disease resistance. The variety Caliverde, developed by E. H. Stanford (16), is resistant to bacterial wilt, mildew, and leafspot because of alleles transferred by backcrossing from several nonrecurrent parent sources. Over 200 plants of the recurrent parent were used in each of 4 backcrosses. Harn (9) cites examples of backcross use in the improvement of several characteristics in oranges and grapefruit, and Anderson (1) suggests the use of backcrossing in clonally propagated perennial crops. In each case the system is refined to satisfy pollination requirements and meet breeding objectives.

Although backcrossing does have some specific advantages as a breeding technique, a serious weakness is involved in its use. Backcrossing, by definition, is a conservative system since genetic variability between the two parents is removed very rapidly. This lowers or eliminates the probability of desirable genetic recombinants for traits other than those specifically programmed in the backcrossing procedure. A modified program, in which only one or two backcrosses are followed by some selection procedure partially overcomes the conservative nature of backcross breeding.

INTROGRESSION

As discussed in Chapter 7, introgression is a system of interspecific crosses followed by repeated backcrosses to one parent species. The backcrossing accomplishes two objectives. It tends to improve the fertility and reproductive capacity of the progeny, and allows recovery of the recurrent parent genotype while including small amounts of genetic information from the nonrecurrent parent. Introgression has been extremely valuable in evolution because of genetic exchange from one species to another. F_1 sterility normally results from an unbalanced cytological situation where chromosomes from each of the contributing species tend to pair very poorly, if at all. The improvement of cytological stability through backcrossing is well demonstrated by Ladizinsky and Fainstein (14) in a study of introgression between cultivated hexaploid oats and two tetraploid wild oat species. Complete self sterility of the pentaploid F_1 hybrids was overcome by massive backpollinations to the parental species. Meiotically stable and reasonable fertile derivatives were selected as early as the F_2 of the BC_1 and from relatively small populations. Gene transfers between the tetraploid and hexaploid species were demonstrated by introducing the allele for nonshattering seed from the cultivated oat to both of the wild species, and lemma hairiness from the tetraploid to the hexaploid. This points out the potential usefulness of introgression in practical plant breeding problems.

Dvorak (6) demonstrated that leaf rust resistant could be transferred through an introgression program from *Aegilops speltoides,* a wild grass species with $2\underline{n} = 14$ to common wheat ($2\underline{n} = 42$). The F_1 hybrids were tetraploid and male sterile. After five backcrosses a large proportion of the progeny were self-fertile hexaploids with normal chromosome pairing, and possessing the disease resistance gene from *A. speltoides.* The resistance genes had been transferred into the wheat chromosomes and were functionally effective in disease reaction.

Hawkes (10) discusses the importance of wild germplasm in plant breeding and gives several examples of valuable gene transfer to commercial crops such as the potato, cucumber, kale, tomato, and others. He points out, however, that sterility problems in this type of breeding program must be overcome through genetic manipulations including backcrossing if maximum use is to be made of this highly valuable genetic variability source.

ISOGENIC LINES

Isogenic lines are groups or sets of material within a variety that differ by only one or a very small number of alleles. Two isogenic lines in a rice variety, for instance, would have exactly the same genetic composition with the exception of alleles governing specific traits such as height or disease resistance. They are derived by backcrossing so that the two allelic forms can be recovered in identical backgrounds from the recurrent parent. For example, if height is governed by a single locus, with tall dominant and short recessive, the alleles can be transferred by backcrossing to any variety. The background genetic purity of the isogenic lines will depend on the number of backcrosses accomplished. True isogenic lines may require 10 to 12 backcrosses while near-isogenic lines may need only 3 or 4 backcrosses.

The value of isogenic lines lies in the ability to determine the effect of particular alleles on the performance and physiology of the plant. For example, it is possible to measure the value of having awns on cereals crops under a specific environment by comparing isogenic lines with and without awns. Such characters as pubesence, disease resistance, responsiveness to chemicals, and different types of plant architecture including leaf size, shape, and color have been investigated.

MULTILINE VARIETIES

A practical breeding application of isogenic lines involves the concept of multiline varieties first suggested by Jensen (12) in oats. We know that the ability of many land race varieties and natural populations to perform well over a wide range of environments is associated with some level of heterogeneity in their composition. As plant breeding programs become more sophisticated and modern agricultural techniques require higher degrees of crop uniformity, the genetic diversity within a variety is destroyed and varieties may become highly vulnerable to many natural hazards such as disease epiphytotics. Each variety is then subject to potential attack following race changes in the disease organism. Borlaug (2) outlined the development of multiline wheat varieties to provide stem rust protection. His program suggested crossing a highly productive recurrent cultivated wheat with nonrecurrent parents having a diverse spectrum of resistance genes. Each derived line would be similar in morphological and performance characteristics, but would contain different resistance genes. The multiline variety would be generated by mixing together the different isogenic backcross lines. Frey et al. (7) describes the development and release of multiline oat varieties with resistance to different crown rust races. Multiline variety E74, for example, contains nine isolines. Nine nonrecurrent parents each contributed a unique and useful gene for rust resistance and five backcrosses were accomplished to the recurrent parent. Over seven years, no increase was observed in susceptibility to crown rust by multiline variety E74. While some disease developed in the crop each year, no serious losses were sustained. This type of protection is the result of several causes. First, there is no severe natural selection for virulent races, since each race is allowed to propagate at a very low level. Second, disease buildup within a field is very slow since many of the spores land on resistant plants. This is particularly important in oats, since the final yield is determined, to a large extent, by seed development very late in the growing season. The use of multiline varieties is still limited, but offers excellent opportunity for pest control management through plant breeding. It also holds promise for the development of varieties buffered against severe environmental fluctuations.

SUMMARY AND COMMENTS

Backcrossing is a conservative but powerful tool in plant improvement by biasing the population in the direction of one parent. It can be a quick, efficient method of responding to problems that can be solved with one or a very few desirable alleles. A high level of backcrossing can be achieved in species that are readily self-pollinated, but cross-pollinated crops must be treated with more caution since inbreeding depression can occur following severe backcrossing. Valuable genes can be transferred from wild species to cultivated plants through introgression. Multiline varieties offer potential adaptation and performance buffering against environmental stress. Isogenic lines have been helpful in understanding the values and functions of different plant morphology features and physiological characteristics.

I have used backcrossing to a limited extent in the investigation of wheat characters associated with dryland production. In particular, we have studied semi-dwarf genes and their relationship to root systems responsible for soil water extraction and crop protection against drought damage. We found that the semi-dwarf genes themselves were not responsible for reduced growth and activity, allowing us to continue using these genes without undue concern for associated drought problems.

I have not used intensive backcrossing to any degree in my wheat breeding program. The conservative nature of this system does not lend itself to the recombination of complex genetic systems controlling important characters in winter wheat produced on dryland. However, I have found a single backcross to an adpated parent helpful on occasion in biasing my segregating populations toward desirable quantitative characters.

REFERENCES

1. Anderson, R. L. 1976. Considerations in backcrossing programs for clonally propagated perennial crops. *Fruit Varieties J.* 30:80.

2. Borlaug, N. E. 1959. *The use of multilineal or composite varieties to control airborne epidemic diseases of self-pollinated crop plants.* In First Int. Wheat Gen. Symp. Proc., Winnipeg, Manitoba. pp. 12-26.

3. Briggs, F. N. 1930. Breeding wheats resistant to bunt by the backcross method. *J. Am. Soc. Agron.* 22:239-244.

4. ———. 1938. The use of the backcross in crop improvement. *Amer. Nat.* 72:285-292.

5. ———, and R. W. Allard. 1953. The current status of the backcross method of plant breeding. *Agron. J.* 45:131-138.

6. Dvorak, J. 1977. Transfer of leaf rust resistance from *Aegilops speltoides* to *Triticum aestivum. Can. J. Gen. Cytol.* 19:133-141.

7. Frey, K. J., J. A. Browning, and M. D. Simons. 1977. Management systems for host resistance genes to control disease loss. pp. 255-274. In P. Day (ed.). *The genetic basis of epidemics in agriculture.* Ann. N. Y. Acad. Sci., 287.

8. Harlan, H. V., and M. N. Pope. 1922. The use and value of backcrosses in small grain breeding. *J. Hered.* 13:319-322.

9. Harn, C. J. 1973. Development of scion cultivars of citrus in Florida. *Proc. Florida State Hort. Sci.* 86:84-88.

10. Hawks, J. G. 1977. The importance of wild germplasm in plant breeding. *Euphytica* 26:615–621.

11. Hough, L. F., J. R. Shay, and D. F. Dayton. 1953. Apple scab resistance from *Malus floribunda* Sieb. *Proc. Am. Soc. Hort. Sci.* 62:341–347.

12. Jensen, N. F. 1952. Intra-varietal diversification in oat breeding. *Agron. J.* 44:30–34.

13. Keep, E., and R. L. Knight. 1968. Use of the black raspberry (*Rubus occidentalis* L.) and other *Rubus* species in breeding red raspberries. Rpt. E. Malling Res. Sta. 1967, 105–107.

14. Ladizinsky, G., and R. Fainstein. 1977. Introgression between the cultivated hexaploid oat *A. sativa* and the tetraploid wild *A. magna* and *A. murphyi*. *Can. J. Genet. Cytol.* 19:59–66.

15. Sprague, G. F. 1955. Corn breeding. pp. 221–292. In G. F. Sprague (ed.), Corn and corn improvement. Academic Press, New York.

16. Stanford, E. H. 1952. *Transfer of resistance to standard varieties.* Proc. Sixth Intern. Grasslands Congr. pp. 1585–1589.

17. Thomas, M. 1952. *Backcrossing, the theory and practice in the breeding of some non-cereal crops.* Commonwealth Agricultural Bureaux, Cambridge.

18. Whitaker, T. W. 1959. An interspecific cross in *Cucurbita* (*C. lundelliana* Bailey × *C. moschata* Duch.). *Madroño* 15:4–13.

14

RECURRENT SELECTION AND SYNTHETIC VARIETIES

Every selection system has as its goal the increase of desirable gene frequency in the population. In self-pollinated species the selection system results in stable homozygous varieties that reproduce faithfully from generation to generation. Even in selfed crops, however, the multiline concept was developed to incorporate heterogeneity for disease protection, suggesting that absolute homogeneity may not be entirely desirable.

In cross-pollinated crops, mass selection was the first type of genetic improvement undertaken. Mass selection was not always effective in producing better yields and, in fact, yields were often depressed as selection continued. This technique worked satisfactorily for high heritability traits such as disease resistance, but did not always achieve success for quantitative traits. Inbreeding depression often occurred when offspring were produced by selfing or by mating of closely related plants. Also, low heritability resulted in lack of ability to identify desirable genotypes based on the phenotypes.

A new approach was needed to maintain natural vigor while concentrating desirable alleles. Recurrent selection was developed as a suitable breeding technique for use in (but not limited to) cross-pollinated crops. This chapter deals with the selection system and its potential application to variety development. In Chapter 15 recurrent selection will be used to identify and improve inbred lines for use in hybrid production.

RECURRENT SELECTION TECHNIQUES

Recurrent selection was first suggested by Jenkins (6) and named by Hull (5). Both were working with corn, as were most of the early researchers interested in this breeding approach.

The basic technique in recurrent selection is the identification of individuals with superior genotypes, and their subsequent intermating to produce a new population. Some type of progeny test, depending on the selection system, may be necessary to measure the genotype value of the parents. Parental genotypes are retained, often by selfing or vegetative propagation, so that, following selection and progeny evaluation, the best ones can be grown and intercrossed. After the intercrossing of superior plants, selection can again be practiced in the new population. The recurring population improvement concept lead to the name "recurrent" selection.

Recurrent selection is cyclic and tends to concentrate desirable alleles while maintaining vigor and genetic recombination through outcrossing. It is distinguished from straight mass selection by the programmed intercrossing of selected individuals and the frequent use of progeny testing to identify superior parents.

GENOTYPE IDENTIFICATION

The key to success in recurrent selection is the correct identification of individuals leading to a population with an increased frequency of desirable alleles. Two identification method alternatives exist. First, individuals carrying desirable genes may be selected by phenotype without progeny evaluation, providing the phenotype accurately reflects the genotype. Highly heritable qualitative traits such as disease resistance and seed size fit this category. In many instances, however, the traits are not easily identified in the individuals to be selected and some type of progeny testing becomes necessary. By measuring the average of a group of offspring, the maternal genotype can be evaluated.

In progeny testing, the gamete source becomes an important consideration. Selfed progeny are produced by conventional methods. S_1 indicates the first generation of selfing, S_2 the second, and so on. They may occasionally be poor indicators of parental genotypic value because of potential inbreeding depression. Testcross progeny are produced by crossing a selected plant as a female with a hybrid or population of individuals containing a known array of genotypes. Topcross, or open-pollinated progeny, are the result of the selected plant being allowed to cross on a random basis with any plants in the area. In both testcross and topcross situations the progeny from a single plant are half-sibs, since the female gametes are from one individual but the male gametes come from a number of different plants.

Progeny tests, regardless of their source, are used as an evaluation system to measure the worth of each parent plant for the next round of intermating in recurrent selection. Each progeny production system is useful in identifying parent value for use in specific breeding programs. Applications of the different progeny tests are discussed in later sections of this chapter and Chapter 15.

CROSSING PROGRAMS TO PRODUCE NEW POPULATIONS

Following selection, parents are placed in a mating situation so that crossing occurs between as many individuals as possible among the select group. This helps maintain vigor and promotes recombination to aid in further concentrating desirable alleles. Several alternatives can be used.

A polycross nursery (Fig. 14.1) is a commonly used system that contains all individuals in a specific physical pattern of arrangement. It must be designed so that each genotype is distributed as randomly as possible with respect to every other genotype. If this is not done, male gametes are not sampled randomly since there is more likelihood of plants mating with adjacent individuals than with individuals some distance away. Polycross nurseries are usually physically isolated from other pollen sources to avoid contamination from unwanted genotypes.

A second option involves the diallel mating scheme in which every individual is manually forced to mate with every other individual. This becomes very laborious as the number of parents increases, but does offer some special opportunities for specific parental combinations and can be considered for limited use in small nurseries. Sometimes only a very specific portion of the diallel (a partial diallel) is used to reduce the number of crosses.

Figure 14.1. A polycross orchardgrass nursery. Each plant is a specific genotype arranged in a geometric pattern. As the plants begin to flower outcrossing will occur in a random manner. (Courtesy C. M. Rinckner, USDA-SEA.)

RECURRENT SELECTION FOR PHENOTYPE

This program selects individuals for characters that can be evaluated phenotypically and is simply a system of selection followed by intermating (Fig. 14.2). The efficiency of the system depends in part on the timing of selection. If desirable genotypes can

Figure 14.2. Recurrent selection for phenotype.

Step I. Grow population and select a series of individual plants for superior phenotype.

Step II. Grow progenies and cross in all possible combinations. Grow a new population of all crossed seed.

Repeat cycle for as long as desired.

be identified prior to flowering, than selection and intermating can be done during the same season. The only restriction is the matching of selected individuals for pollination. If this is a problem in the field, it is possible to take the plants, in the case of cloned species, to a greenhouse and complete the crossing program there.

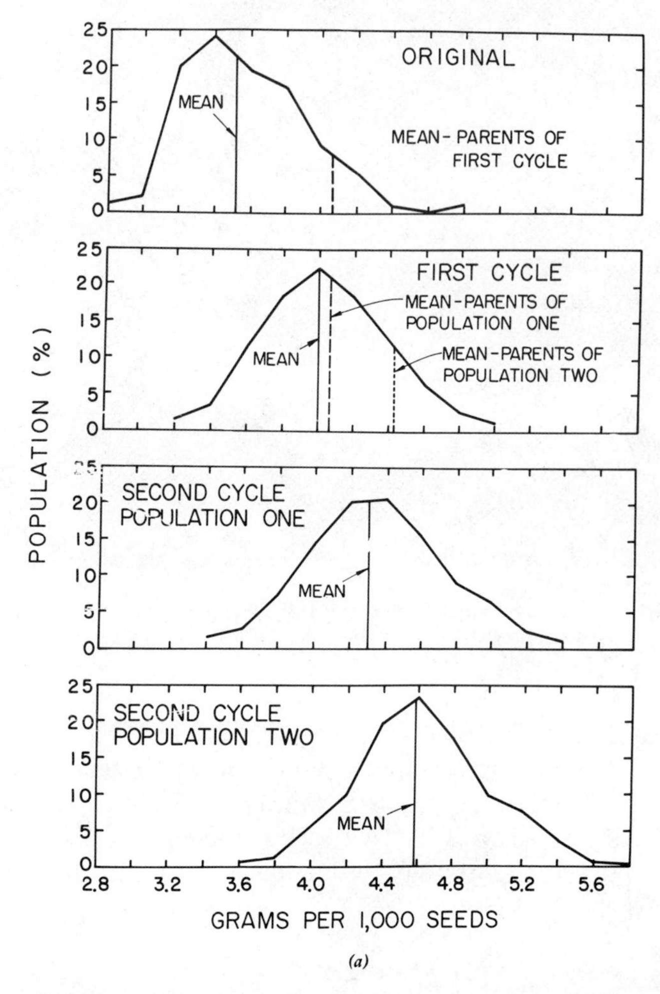

(a)

Figure 14.3. Recurrent selection for seed weight in cicer milkvetch using two mating systems. In both mating systems the original population was the same. In subsequent selection cycles within each mating system, population one was selected for both vigor and seed weight while population two was selected primarily for seed weight. (*a*) diallel cross mating system. (*b*) polycross mating system. (From *Crop Science,* 1977, 17(3):474, by permission of the Crop Science Society of America.)

If the phenotype cannot be identified until after pollination, than either selfed or crossed seed must be produced on each selected individual. Progress, of course, is slower with crossed seed than selfed seed because in crossed seed the male gametes may come from either unknown or undesirable genotypes.

After selection and crossing has been completed the cycle is repeated by planting the F_1's and selecting phenotypically desirable individuals from this new population. Often, two rounds of recurrent selection for phenotype are sufficient to exploit the ge-

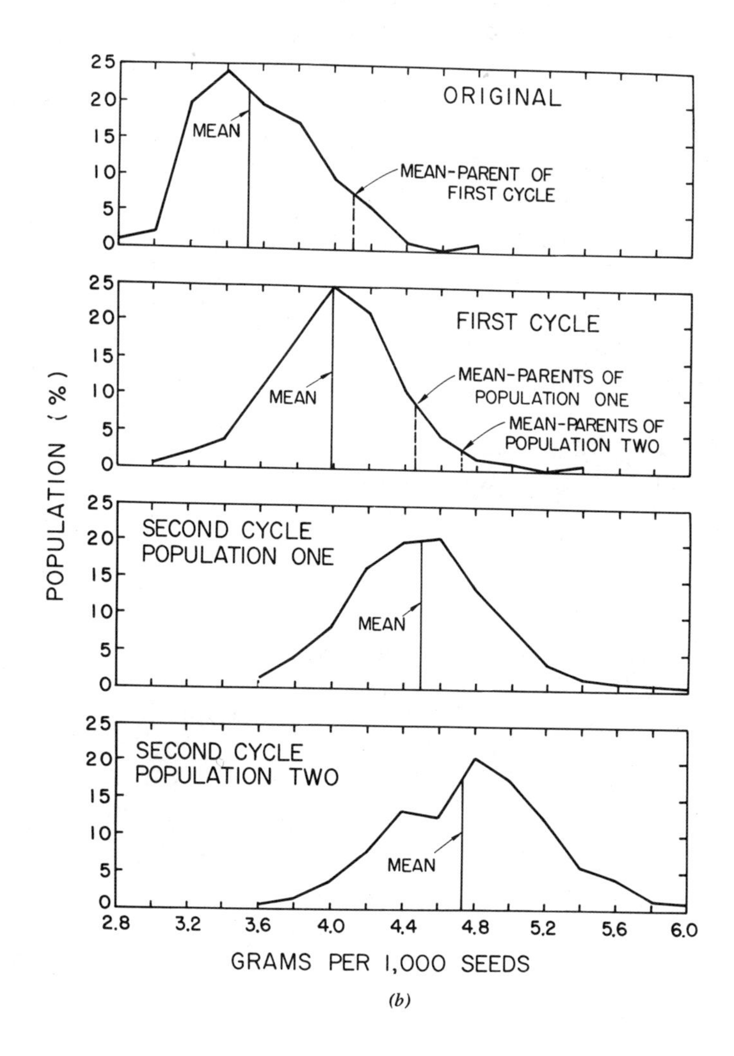

(b)

netic variability present, and the system becomes inefficient for gains from further selection. The number of effective cycles is based on the types of characters selected and the amount of available genetic variability.

Townsend (10) provides an example of improvement achieved through recurrent selection for phenotype with the character of seed weight in cicer milkvetch. Results are given in Figure 14.3. Emergence and stand establishment are important characteristics in this legume crop and depend on seed weight and plant vigor. Two systems of intermating—diallel and polycross—were compared. In the first cycle, plants with good vigor and high seed weight were used for intercrossing. The progeny from this cycle showed a 13 percent seed weight increase. In the second cycle two populations were utilized. Population one was heavily selected for plant vigor with reduced emphasis on seed size. In population two, vigor was deemphasized, but seed weight received increased selection pressure. The results of the second cycle show that seed weight had increased between 22.5 percent and 34.8 percent over the mean of the original populations. The greatest seed weight was achieved in the polycross system where plant vigor received only a small amount of emphasis. This indicates that a single selection objective will result in the greatest gain and may be obtained at the expense of gains in other characters. Diallel matings reduced population variation more rapidly than the polycross system, but this may have resulted because a smaller number of parents were used. While good gains were made in seed size, the extent of remaining variability could not be determined accurately without further selection studies.

Recurrent selection for phenotype has been carried out successfully in the cross-fertilized grasses, forages (especially alfalfa), and corn. This same type of selection system can also be conducted in self-fertilized crops. Boomstra and Bliss (1) studied the effect of recurrent selection for resistance to root rot in beans. The resistance appears to be quantitative in nature and does not lend itself readily to pedigree or backcross methods. They suggest that recurrent selection is the best alternative in concentrating the desirable resistance genes in a single population. Recurrent selection for phenotypic improvement has not been used extensively in self-pollinated crops, however, because of mechanical limitations. Compton (3), in considering recurrent selection in selfed crops, suggests using the single seed descent system following F_1 production, as an aid in reducing the labor requirements. Each F_1 from the diallel cross of a set of selected parents is propagated on a single seed basis to the F_4 or F_5 for yield testing, the top selections are intermated, and the procedure repeated. The key is the single seed needed for the F_1 of each cross that reduces the effort required while concentrating desired genes through selection. In Chapter 16 we introduce a sterility concept that could be an aid in the utilization of recurrent selection in self-pollinated crops.

RECURRENT SELECTION FOR SPECIFIC COMBINING ABILITY

In some cases, especially in hybrid breeding, it is desirable to have breeding lines that produce a high proportion of productive progeny when combined with other narrow genetic base stocks. This is called specific combining ability (SCA) and is considered to include both additive and nonadditive types of gene action. Recurrent selection, as outlined in Figure 14.4, can be applied to this character. The tester in this case must be a line, such as a specific inbred, that has limited genetic diversity. Evaluation of the selected plants is done by measuring average performance of the progeny. Reserve seed of the best female parents are planted, the plants intercrossed, and the cycle repeated. Since specific combining ability is a primary concern in hybrids, Chapter 15 will cover this subject in more detail.

Figure 14.4. Recurrent selection for specific combining ability.

Step I. Self-pollinate several individuals and also cross to narrow genetic base tester. Save selfed seed in reserve.

Step II. Grow testcross progenies and evaluate for yield. Identify original individuals with superior specific combining ability.

Step III. Grow reserve selfed seed of each superior original plant and intercross.

Repeat cycle for as long as desired.

RECURRENT SELECTION FOR GENERAL COMBINING ABILITY

Often, we are interested in the identification of parents in a cross-pollinated crop that, when combined with several other genotypes, will produce a high proportion of very vigorous, productive offspring. These individuals are said to have good general combining ability (GCA) and are useful in the development of synthetic varieties and hybrids. The expression of general combining ability relies heavily on additive gene action. Recurrent selection (Fig. 14.5) can be used for improvement in general combining ability. Measurement of GCA is done by progeny testing. The male tester must provide a diverse set of gametes and is often an open-pollinated variety or the polycross nursery itself. The seed harvested on the individual in question represents a wide assortment of male gametes. If the average performance of the progeny is very good, than the female is considered to have good GCA since she combines well with many genotypes. Following the identification of individuals with high GCA, the remnant seed is planted in isolation and crossed to concentrate the alleles responsible for combining ability. Since general combining ability is likely the result of many loci, improvement progress may be slow. In many cases selection pressure is applied for GCA and phenotype at the same time to maintain vigor while concentrating specific valuable alleles.

RECIPROCAL RECURRENT SELECTION

Reciprocal recurrent selection improves both specific and general combining ability. Two populations are selected simulta-

Figure 14.5. Recurrent selection for general combining ability.

Step I. Self-pollinate several individuals, and also cross with a wide genetic base tester. Save selfed seed in reserve.

Step II. Grow testcross progenies and evaluate. Identify individuals with superior general combining ability.

Step III. Grow reserve selfed seed of each superior original plant and intercross.

Repeat cycle for as long as desired.

Figure 14.6. Reciprocal recurrent selection.

Step I. Plant two different populations. In each population self several selected individuals and also outcross them to the other population. Save selfed seed in reserve.

Step II. Grow testcross progenies and evaluate. Identify original individuals with superior combining ability.

Step III. Grow reserve selfed seed of each superior original plant, keeping populations separate. Intercross within each population.

Repeat cycle for as long as desired.

neously with each population serving as the tester for the other. See Figure 14.6. Selected individuals within each population are crossed with a random sample of plants from the other population, and the progeny are tested. The best of the selections are replanted and intercrossed, completing one cycle. In the second cycle, selected plants in each group are again crossed with the other population as the tester. This carries with it the implication that the genetic composition of each tester changes with each cycle of selection. Jugenheimer (7) and Sprague and Eberhart (9) summarize research in reciprocal recurrent selection with corn and show that it has been used with varying degrees of effectiveness for several traits. This selection system has the advantage of capitalizing on nonadditive gene effects and simultaneous improvement of two base populations.

SYNTHETIC VARIETIES

A synthetic results from natural intercrossing of two or more strains or clones, which are thus "synthesized" into a new variety. The crossed seed is harvested in bulk and replanted in successive generations. The parents are the SYN 0, the seed from the first round of mating the SYN 1 (first generation synthetic), seed harvested from the SYN 1 is SYN 2, and so on. In its strictest definition a synthetic variety is the result of random mating after any round of recurrent selection, but is normally reserved for varieties that are put on the commercial market. Hays and Berger (4) first introduced the term in corn in 1919. Parents for a synthetic are a group of individuals selected for some set of traits. They are allowed to intermate in much the same manner as in open-pollinated varieties. Synthetic varieties are produced as the result of phenotypic selection for certain traits along with good vigor level measured by general combining ability of selected lines.

The first corn synthetic was described by Sprague (8) in 1946. A group of 16 lines, called "stiff stalked" were selected for resistance to lodging and combined to produce the "stiff stalked synthetic." This synthetic has not only proven to be a valuable variety of corn, but it is also a useful source of germplasm for recurrent selection in many breeding programs.

After the breeder has selected the lines to be included in the SYN 0, what can be expected in yield performance after each round of mating? If we assume Hardy-Weinberg population dynamics, each individual will have an equal opportunity to mate with every other individual and, unless some very restrictive mating mechanism is present, some selfing will occur. The normal expectation, proven by extensive experimental data, is that the SYN 1 will have the highest yield performance because of maximum genetic diversity and hybrid vigor expression. The SYN 2 generation yield is normally somewhat reduced due to the production of some homozygous genotypes leading to limited inbreeding depression and reduced yield. After the SYN 2, yield appears to stabilize and very little reduction in vigor occurs. This is in agreement with the Hardy-Weinberg principle of stabilized gene and genotype frequencies after one round of random mating.

Difficulties arise, however, because of natural selection that can change successive generation genotypes of the synthetic. In alfalfa, for example, seed harvested from a SYN 2 field that has experienced partial winterkill will have a different population genotype and gene frequency than the SYN 1. Also, genetic shifts by nonramdom mating and chance alone can take place so that some genotypes are lost. When these conditions occur, the synthetic gradually changes over time and may be reduced in yield performance. Thus, it must be reconstructed periodically, requiring the breeder to maintain all the original lines used in its development. The number of lines and the proportion of each line to be included in the synthetic depends partly on the amount of inbreeding depression experienced in the species. In alfalfa, severe inbreeding depression occurs with mild levels of selfing or of closely related matings. Experimental evidence suggests up to 16 lines are desirable in the composition of a synthetic for maintenance of vigor and productivity. The number varies considerably with each species and breeding program, however.

Synthetic varieties provide a mechanism to capitalize on phenotypic selection while maintaining vigor. They also allow the grower to reproduce seed for a few generations without entirely sacrificing genetic integrity of the variety. Snythetics are used extensively in most legumes and grasses. Busbice et al. (2) indicates that nearly all alfalfa varieties sold today are synthetics. They are also common in corn where production is marginal, and where hybrid corn is not feasible because of seed production and distribution or economic limitations. In many developing countries synthetic corn varieties offer the possibility for using a breeding system that concentrates desirable genes while maintaining high levels of natural diversity.

Synthetic varieties have also been used as a maintenance base for genetic diversity. An excellent example is provided by CIMMYT where corn populations have been selected for characters such as disease resistance, while retaining natural vigor and genetic variability for most other traits. The populations are distributed to breeders around the world for them to impose whatever selection pressure fits their local needs. This provides valuable genes to many programs with limited resources.

SUMMARY AND COMMENTS

Recurrent selection is a powerful breeding tool, especially in open-pollinated species where natural vigor must be maintained through crossing. Synthetic varieties represent a practical application of recurrent selection in which hybrid vigor is utilized, but with the added advantage of potential propagation by the producer for several generations.

Although I have not used a specifically designed recurrent selection program in my wheat work, it does appear in modified form each time I recycle some of the better advanced lines through the crossing program.

This, in effect, tends to concentrate desirable alleles while promoting recombination for many others. The value and use of recurrent selection in selfed species is yet to be exploited and represents another breeding challenge for the future.

REFERENCES

1. Boomstra, A. G., and F. A. Bliss. 1977. Inheritance of resistance to *Fusarium solani* f. sp. *Phaseoli* in beans (*Phaseolus vulgaris* L.) and breeding strategy to transfer resistance. *J. Amer. Hort. Sci.* 102:186–187.

2. Busbice, T. H., R. R. Hill, Jr., and H. L. Garnahan. 1972. Genetics and breeding procedures. pp. 283–318. In C. H. Hanson (ed.). *Alfalfa science and technology.* Amer. Soc. Agron., Madison, Wisc.

3. Compton, W. A. 1968. Recurrent selection in self-pollinated crops without extensive crossing. *Crop Sci.* 8:173.

4. Hays, H. K., and R. J. Gerber. 1919. Synthetic production of high-protein corn in relation to breeding. *J. Am. Soc. Agron.* 11:309–318.

5. Hull, F. H. 1945. Recurrent selection for specific combining ability in corn. *J. Am. Soc. Agron.* 37:134–145.

6. Jenkins, M. T. 1940. The segregation of genes affecting yield of grain in maize. *J. Am. Soc. Agron.* 32:55–63.

7. Jugenheimer, R. W. 1976. *Corn improvement, seed production and uses.* Wiley, New York.

8. Sprague, G. F. 1946. Early testing of inbred lines of corn. *J. Am. Soc. Agron.* 38:108–117.

9. ———, and S. A. Eberhart. 1977. Corn breeding. pp. 305–362. In G. F. Sprague (ed.). *Corn and corn improvement.* Am. Soc. Agron., Madison, Wisc.

10. Townsend, C. E. 1977. Recurrent selection for high seed weight in cicer milkvetch. *Crop Sci.* 17:473–476.

15

HYBRID BREEDING

Revolutionary advances have taken place in plant breeding techniques and methods since the rediscovery of Mendel's laws in 1900. Of all the breeding innovations that have been developed, one of the most fascinating and productive has been the hybrid system.

Researchers and breeders regularly observed that F_1's often had an increased level of vigor and performance when compared with the parents. As early as the 1700s, heterosis in the form of improved performance and vigor resulting from crossing was described for several species. Likewise, the inbreeding of naturally cross-fertilized plants often resulted in decreased vigor and the appearance of abnormal individuals. An interesting historical summary on inbreeding and heterosis is edited by Gowan (5).

The development of genetic concepts and principles associated with corn inbreeding and heterosis became economically important in the 1930s. Following this, a rapid expansion and refinement occurred in hybrid breeding and the seed industries of many species. Corn has received the bulk of attention in hybrid research and much of the discussion centers around that crop. The student is encouraged to remember that the same principles have been applied equally well to many other plant and animal species.

INBREEDING DEPRESSION

Inbreeding depression or vigor loss following the mating of closely related individuals has been mentioned in previous chapters. The documentation of this phenomenon has taken place over many hundreds of years. It was most evident in animal species, particularly humans, where detrimental effects began to appear following the mating of close relatives. East (2,3) and Shull (10) first documented and described inbreeding depression in plants. They observed that selfing individuals in a naturally cross-pollinated corn population almost invariably reduced vigor in the progeny, especially when selfing was accomplished for several generations. In fact, vigor loss was so severe in some cases that progeny propagation was impossible. Other inbred lines experienced some depression, but later stabilized in performance.

The cause of inbreeding depression is likely associated with the uncovering of deleterious recessives and lethals in homozygous genotypes. In addition, any advantage of vigor from heterozygous genotypes is diminished with selfing. Inbreeding depression does not normally occur in naturally selfed plants since a high level of homozygosity already exists.

HETEROSIS

Heterosis definitions differ depending on the base of comparison used. In this discussion we will define heterosis as the improvement of the F_1 over the best parent. Other definitions include the comparison of the F_1 with the mean of the parents, with the nursery average, or with the average of a set of check varieties. In cases of F_1 performance less than midway between the two parents or poorer than the worst parent, negative heterosis is implied.

Shull (11) pointed out that certain crosses of corn inbreds produced remarkable growth and vigor in their progeny while the expression of heterosis in other crosses was either very small or nonexistent. He was thus implying that useable economic heterosis was available following the crossing of certain (but not all) corn inbred lines. This suggestion is the foundation of the hybrid plant and animal industry since the secret of success in the hybrid business lies in the ability to find the correct combination of inbreds to produce competitive hybrids for the marketplace.

The genetic causes of heterosis are not completely understood, but several possible explanations have been advanced. It was originally thought to be the favorable expression of heterozygosity. In fact, heterosis is a contraction of the word heterozygosis. One of the most popular theories has been that of favorable enzymatic production associated with heterozygous alleles. That is, a hybrid enzyme system driving the plant mechanism is produced by heterozygous allelic forms. This theory has been difficult to demonstrate, but is continuing to receive attention and some evidence is accumulating that it may partly account for heterosis.

Another possible explanation is the accumulation of favorable dominant alleles in the F_1, some of which are contributed by each parent. Detrimental effects caused by homozygous recessive alleles would be masked over. If this is the cause, then inbred lines containing all homozygous dominant alleles in their genotypes should be possible through breeding and selection, and heterosis would be permanently fixed as a true breeding condition. While this has not yet been achieved, inbred lines are continually being improved genetically. In addition, linkage may make the accumulation of all dominants a very difficult task since correct recombination must take place between every dominant and recessive allelic combination. Considering the number of loci, the probability of this happening with the subsequent selection of correct gametes may be extremely low.

A third potential possibility lies with the cytoplasm and its relationship to chromosome information. In this explanation at least part of the heterotic performance is due to material contained outside the nucleus. Hereditary information is carried in some extranuclear organelles such as the mitochrondia. There is some apparent interaction between the chromosomes and the cytoplasm, as demonstrated by some forms of male sterility. Different nuclei can be placed in a cytoplasm by backcrossing and this technique provides some evidence to support the concept of a nucleus-cytoplasm effect in heterosis. Results, however, are not clear-cut. The true explanation probably includes all three theories.

A partial list of plants in which heterosis has been evaluated include corn, sorghum, wheat, oats, barley, rice, sugar beets, alfalfa, tomato, spinach, cabbage, broccoli, cauliflower, kale, carrots, cucumbers, sunflowers, petunias, marigolds, begonias and geraniums. The earliest cases of heterosis were demonstrated in naturally cross-pollinated crops where inbred lines had to be developed.

Self-pollinated species provided a wealth of ready-made inbred lines without manipulation by the breeder. Hayes and Foster (4) point out that while heterosis has been demonstrated in many naturally selfed crops, it is often not as pronounced as in crossed species. Possibly the self-pollinated crops have fixed heterosis in a homozygous condition over the years through natural selection for the most productive genotypes.

In both crossed and selfed species, however, enough heterosis exists in certain crosses to justify research and development input for potential economic competitiveness in the marketplace. To make hybrids a commercial possibility, two problems must be addressed. These include (1) the identification of highly productive hybrid combinations and (2) the production of hybrid seed on an economic basis.

COMBINING ABILITY

Combining ability, introduced in Chapter 14, is defined as the ability to produce desirable progeny. It might be logical to assume that the greatest heterosis would be displayed by crosses of pure line inbreds experiencing the most inbreeding depression. In actuality, however, heterosis dependent on combining ability can occur in any combination of pure line inbreds. In wheat, for example, some of the greatest heterosis occurs in hybrids of very productive pure lines.

General combining ability (GCA) is expressed in the progeny of an inbred crossed with many genotypes, and is primarily the result of additive gene action. Specific combining ability (SCA) is the expression of performance between any two inbred lines and is attributed to dominant, epistatic, and additive gene action. Both types of combining ability are important in the identification of valuable inbred lines for use in hybrids.

To test for GCA, the inbred line is crossed with a spectrum of genotypes such as open-pollinated corn varieties, all individuals in a polycross block, or some wide genetic based hybrid. Following crossing with the tester, the average value of the progeny is measured.

Specific combining ability is very important in hybrid breeding since financial success depends on identifying the few very superior combinations. SCA values are obtained either by crossing all inbreds in a diallel manner or by developing some prediction system to allow estimation of performance without actually making all crosses. As more than a few inbreds become available for testing, the prediction system takes on added importance because of the rapid increase in hybrid combination numbers. For example, 10 inbred lines will produce 45 unique single crosses, 340 unique three-way crosses, and 630 unique double crosses, not including reciprocals. Three-way crosses are those in which an F_1 of two inbreds is crossed with a third inbred, and a double cross is produced by the crossing of two F_1's each from two inbred lines. The numbers rapidly become unmanageable when one actually evaluates all combinations, especially in the case of double cross hybrids.

Several methods of estimation have been developed and are summarized by Sprague and Eberhart (12) and Jugenheimer (7). The best estimator for the double cross (A × B)(C × D) appears to be the average of the four single crosses (A × C), (A × D), (B × C), and (B × D) where each capitol letter identifies an inbred line. Note that the combinations (A × B) and (C × D), which are the single crosses actually used to produce the double crosses, are not included in the average. In the three-way cross of (A × B) × C the best predictor would be the single crosses (A × C) and (B × C), again omitting the specific F_1 of (A × B) which was used in producing the three-way cross with inbred C. This estimation system allows the breeder to evaluate potential three-way and double cross hybrid performances without actually using the labor and the time of at least two years to produce the hybrids for measurement. By making the 45 single crosses with 10 inbreds, 340 three-way and 630 double crosses can be estimated. As a group of potential top performers are identified, the crosses are made to obtain true production values for final distribution and sale decisions. This is not the only prediction system available, but it does

Figure 15.1. A hybrid sorghum breeding nursery. The white bags are handmade crosses to produce new hybrids. In this program more than 10,000 hybrids are screened and tested annually. (Courtesy DeKalb AgResearch, Inc.)

provide an idea of the technique used in handling large numbers of inbred lines. The identification of combining ability is a major effort in any hybrid breeding program. An illustration of a hybrid sorghum breeding nursery is given in Figure 15.1

A recent innovation has been suggested as a way to reduce the time and expense associated with field measurement of hybrid performance. McDaniel and Sarkissian (8) proposed measuring the physiological activity of mitochrondia from hybrids in the seedling stage. They produced evidence that mitochondrial activity from hybrids was considerably elevated over that of the inbred parents in both corn and barley, and suggested that mitochondrial complementation could be used as a predictor of heterosis in hybrids without growing the plants to maturity. Barattt and Flavell (1) in a 1977 review state that mitochondrial complementation has been studied in an array of crops including sugar beets, wheat, corn, and barely, and the expression of heterosis in mitochondrial activity can be measured in some cases but the relationship between activity and yield performance has been difficult to demonstrate in a large number of hybrids. Since variation does exist in this biochemical system, however, further research is necessary to determine the potential for use in hybrid performance prediction.

POPULATION AND INBRED LINE IMPROVEMENT

Genetic development and improvement of inbred lines is imperative in any hybrid breeding system. The first type of corn population genetic improvement utilized mass selection with ear-to-row modifications. Mass selection was simply the selection of a group of individuals bulked for the next generation. Ear-to-row selection was the maintenance of an individual row from each plant selected with no control over the pollen source. Since the spectrum of genetic diversity decreased with each generation of selection, some inbreeding automatically occurred. During the inbreeding process, visually identifiable specific phenotypic characters including stem strength and disease and insect resistance were selected. While the inbreds could be improved for specific properties, the mass selection type of improvement program did not allow the breeder to evaluate the potential of the inbreds in hybrids. It was thus possible to put a tremendous amount of effort into inbred improvement without having any real assurance that more productive hybrids would result.

Early generation testing, before a high degree of selfing and homozygosity had occurred, seemed to be appropriate in providing initial clues about the potential usability of populations in hybrids and to allow for the improvement of population combining ability. Three types of recurrent selection systems, described in Chapter 15, have been used to accomplish this objective. The first of these—recurrent selection for general combining ability—evaluates and improves the ability of the inbred to combine with a wide array of genotypes. The tester used to produce the progeny for evaluation would be an open-pollinated variety or wide genetic base hybrid. The populations identified and improved for general combining ability often will have a higher probability of good specific combining ability also. In addition, this allows for the development of synthetic varieties that may be usable in areas not feasible for hybrid production.

Recurrent selection for specific combining ability is also used in inbred line improvement. Here, individuals are evaluated by crossing with one or a number of highly inbred lines to select for high specific combining ability.

Reciprocal recurrent selection is the third system used for inbred improvement and evaluation. In this program two populations are improved at the same time, with each population being used as the tester source for the other. Selection pressure can be applied for any characters desired, but in most cases general combining ability is the basis of improvement. Reciprocal recurrent selection improves efficiency in the breeding program since the tester crosses can serve as a selection improvement base for both populations simultaneously.

Standard pure line development programs, such as the pedigree system, can be used to produce potential parents for hybird combinations. This has been used in several crops such as wheat, barley, and sorghum where the primary thrust of breeding programs was pure line variety release. However, the breeding of pure lines simply makes them available for use in any variety or hybrid system provided the end use results in increased productivity and utility.

As with all other breeding programs, genetic diversity must be introduced periodically into the population to provide new recombinations and selection potential. This is accomplished by outcrossing to other populations, by using exotic sources of variation, and by backcrossing to incorporate specific genes of high value into the inbreds. Here the approaches are very similar to those in conventional nonhybrid breeding programs. Parallel systems are normally operated in all hybrid breeding, where inbreds are improved concurrently with the actual development of the hybrids themselves.

HYBRIDS

Although it was demonstrated in the early 1900s that heterosis would result from the crossing of some corn inbreds, the actual development of the hybrid corn industry did not occur until the mid 1930s. The reason for this was the low seed volume produced when one inbred was crossed with another, a reflection of the reduced vigor of the inbred lines. The small amount of hybrid seed required a prohibitive seed price to recover production costs. Jones (6) in 1922 suggested the use of the double or four-way cross, in hybrid corn production. In this system two very vigorous F_1's from specific inbred combinations are crossed together to produce the hybrid seed sold to the farmer. Because of the heterosis associated with each F_1, the amount of hybrid seed produced was greatly increased per unit of economic input and it was then possible to provide hybrid seed to the producer at an affordable price. After this development the hybrid corn industry expanded rapidly until it now occupies practically all the production acreage in the United States. With the genetic improvement of vigor in inbred lines, hybrid companies have been able to make the transition to three-way and single cross hybrids that currently dominate the hybrid acreage.

Following the pioneering effort in corn, hybrid breeding was applied to many crops. Hybrids are now used extensively in sorghum, sugar beets, onions and many other vegetables, and a large group of ornamental flowers (9). Commercial organizations play a major role in the development and sale of hybrids because of the built-in protection provided by inbred line ownership. Producers on occasion have attempted to use the F_2 generation as a crop the following year without returning to the company for seed. The result is a wide array of phenotypes in the production field, since the F_2 is segregating for any characteristics differing between the inbreds. Modern production and harvesting techniques do not work well in populations containing a divergence of genotypes. Because of this, the producer must buy new hybrid seed each year. This amounts to controlling seed sales by virtue of owning the inbred lines used to produce the hybrid. Since commercial organizations are profit motivated, the hybrid seed industry offers sound economic possibilities for investment in research and development.

Basic information in hybrids has primarily been generated by public agencies. Following the establishment of hybrid possibilities in a particular crop, industry has completed the economic framework of development and sales. This joint effort represents an excellent example of cooperation between public and private programs, and has contributed greatly to the development of productive, efficient agriculture in the United States and throughout the world.

Hybrid systems are still being developed in many crops, especially those that are self-pollinated. Special challenges exist in these crops and unique mechanisms must be introduced genetically. These are discussed in Chapter 16.

SUMMARY AND COMMENTS

Hybrid vigor has been demonstrated and exploited in many plant species. Principles of heterosis are applicable to self-pollinated as well as cross-pollinated plants and pure line breeding programs are a rich source of inbreds for hybrids. Commercial competitiveness in the hybrid seed industry requires breeders to continually improve and update breeding methods for inbred line improvement and identification of combining ability.

I believe that genetic information and breeding systems in ornamental and agricultural species have been stimulated extensively by the hybrid industry. In the United States there would probably be very little commercial involvement in the seed business without hybrids. It would be incorrect, however, to assume that hybrids are the genetic panacea of all agricultural production problems. In many places throughout the world hybrids are simply not feasible because of limited financial resources on the part of the producer, and lack of a sophisticated seed industry.

Breeders have done a magnificent job of improving the genetic potential and properties of inbred lines in corn, to the point where they are becoming quite productive in their own right. The arguement is often put forth that if the amount of effort devoted to the hybrid industry had been put into the improvement of open-pollinated varieties, approximately the same production potential may have been achieved. I don't believe that this is true and feel that there is some distinct production advantage to the hybrid condition itself. The two types of breeding programs are gradually converging, however, and it will be interesting to watch the next thirty or forty years of hybrid breeding improvement.

REFERENCES

1. Barratt, D. H. P., and R. B. Flavell. 1977. Mitochondrial complementation and grain yield in hybrid wheat. *Ann. Bot.* 41:1333–1343.

2. East, E. M. 1908. *Inbreeding in corn.* Annual Rep. Conn. Agric. Exp. Sta. for 1907. pp. 419–428.

3. ———. 1909. The distinction between development and heredity in inbreeding. *Am. Nat.* 43:173–181.

4. Hayes, J. D., and C. A. Foster. 1976. Heterosis in self-pollinated crops, with particular reference to barley. pp. 239–256. In A. Janossy and F. G. H. Lupton (eds.). *Heterosis in plant breeding.* Elsevier Scientific, New York.

5. *Heterosis.* 1964. J. W. Gowen (ed.). Hafner, New York.

6. Jones, D. F. 1922. The productiveness of single and double-cross first generation hybrids. *J. Am. Soc. Agron.* 14:241–252.

7. Jugenheimer, R. W. 1976. *Corn improvement, seed production and uses.* Wiley, New York.

8. McDaniel, R. G., and Sarkissian, I. V. 1966. Heterosis: complementation by mitochondria. *Science N. Y.* 152:1640–1642.

9. Reimann-Phillips, R. 1976. Breeding of F_1-hybrids in flowers. pp. 135–146. In A. Janossy and F. G. H Lupton (eds). *Heterosis in plant breeding.* Elsevier Scientific, New York.

10. Shull, G. H. 1908. The composition of a field of maize. *Am. Breed. Assoc. Rep.* 4:296–301.

11. ———. 1909. A pure line method of corn breeding. *Am. Breed. Assoc. Rep.* 5:51–59.

12. Sprague, G. F., and S. A Eberhart. 1977. Corn breeding. pp. 305–362. In G. F. Sprague (ed.). *Corn and corn improvement.* Am Soc. Agron., Madison, Wisc.

16

HYBRID PRODUCTION SYSTEMS

Following the demonstration of usable economic levels of heterosis in hybrids, the science and technology of hybrid seed production expanded rapidly. The hybrid seed business is finely tuned economically since seed must be sold at a price that will give a reasonable return to the company, while not being prohibitive for the purchaser. This chapter is devoted to hybrid seed production systems and mechanisms.

HANDMADE HYBRIDS

In corn, hybrids can be made manually. The monoecious morphology allows easy emasculation prior to anthesis simply by pulling or breaking the tassel stalk. Detasseling commonly requires the employment of considerable summer labor, primarily students, although it can be done with special mechanical equipment. Seed produced on detasseled rows is the result of the female gametes in the ear and the male gametes from the pollen source provided. By harvesting the detasseled rows and maintaining the seed identity, hybrids can easily be produced in large volume. The major requirements of this system are isolation so that foreign pollen from unwanted male sources is not available to fertilize the detasseled females, and a thorough job of detasseling with minimum plant damage. Isolation is done either through spatial distance or through the planting of surrounding areas with the male, maintaining a pure pollen source for fertilization. The detasseling job is under the direction of a production supervisor who very carefully oversees the labor force and the mechanical equipment.

Since all plants do not tassel on the same day, several operations may be necessary to complete the job. If fields are wet at tasseling time, it may be impossible to move machinery—in which case all detasseling must be done by hand.

The normal practice in hybrid seed corn production as reviewed by Craig (1) is to plant several rows of a female, alternated with a number of male rows, followed again by several female rows, and so on (Fig. 16.1). This results in a field where the females can be harvested separately and the hybrid seed kept pure. Male rows are sometimes destroyed by cultivation or stem breakage after pollination.

Labor economics has prompted the search for less expensive ways to produce hybrid seed. Genetic and cytoplasmic male sterility mechanisms manipulated through the seed source have become increasingly attractive in cross-pollinated crops and are mandatory in large volume selfed crops such as sorghum, wheat, barley, and rice to reduce seed costs. In low volume high economic return

Figure 16.1. A hybrid seed corn production field using a 6:2 pattern. Six rows of detasseled females are alternated with two rows of a pollinator. (Courtesy M. Zuber, Univ. of Missouri.)

species, such as tomatoes and most of the ornamentals, hand pollination is still used. Many of these species are grown by home gardeners who are willing to obtain the desirability of hybrids with a slight additional expenditure simply because their seed requirements are so low. In all plant species where hybrids are being developed, however, the search continues for mechanisms to eliminate high cost labor inputs. Duvick (4) and Frankel and Galun (5) review a number of potential systems to alter breeding methodology.

CYTOPLASMIC MALE STERILITY

Since hybrids are the result of controlled pollinations, some inherited system to render the male nonfunctional would be highly desirable. In addition, this system should be reversible on demand so male gametes could be produced when needed. Remember, the economic worth of hybrids in some species is not dependent on seed production, in which case male gametes are not needed. In onions, for example, the F_1 crop is the bulb, so seed production is not necessary and, in fact, may be undesirable. Likewise, in sugar beets the F_1 economic crop is the root. In many ornamental species such as the petunia, the important organ is the flower itself. On the other hand, the worth of agronomic crops such as corn and sorghum is dependent in part on the amount of seed produced, while some, such as the cereals, are grown exclusively for their grain production.

Cytoplasmic male sterility is the most widely utilized system in the hybrid industry. Every cell is made up of two general components, the nucleus that contains, among other things, the chromosomal information, and the cytoplasm that comprises the remainder of the cell. Some inheritance mechanisms are carried in the cytoplasm (6). In fertilization the female contributes a haploid nucleus and virtually all the cytoplasm through the egg, while the male contributes a haploid nucleus but almost no cytoplasm through the pollen. This results in the female cytoplasm being passed from generation to generation.

If a cross ♀ A × ♂ B is made, the zygote nucleus will be 50 percent A and 50 percent B, while the cytoplasm will be almost entirely A. The reciprocal cross would have the same nuclear constitution, but with B cytoplasm. Information carried in the cytoplasm (10) affecting the phenotype will be contributed exclusively by the female. A specific cytoplasm can be carried along from generation to generation, provided the individual possessing it is used as the female in each cross. A backcrossing program can be carried out so that other nuclei can be systematically substituted into a given cytoplasm. See Figure 16.2.

The interaction of genes and cytoplasm on male gamete production was demonstrated in the late 1930s in onions. A recessive

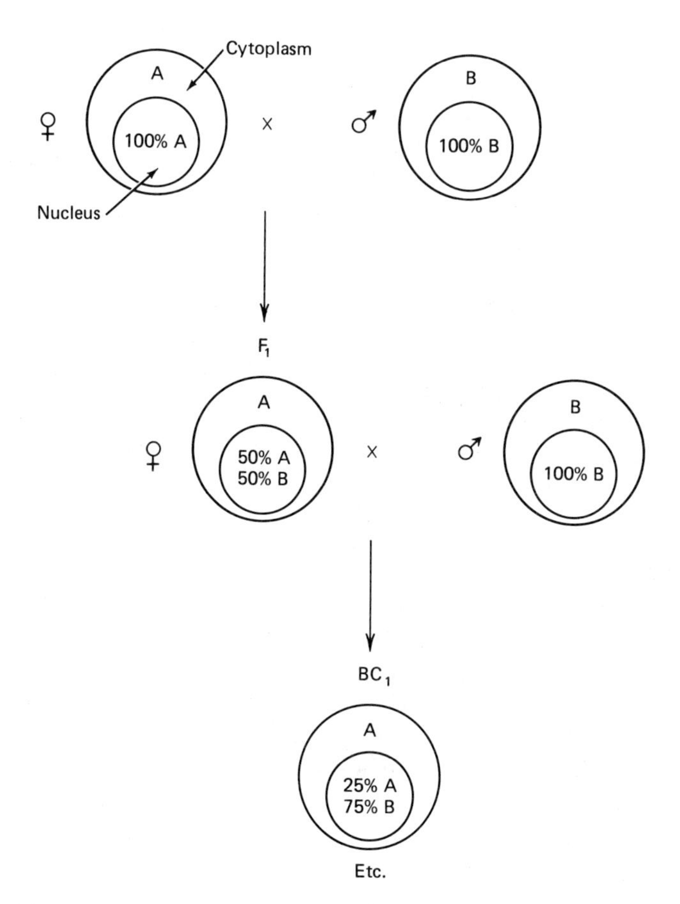

Figure 16.2. A backcrossing program to substitute a different nucleus into a cytoplasm.

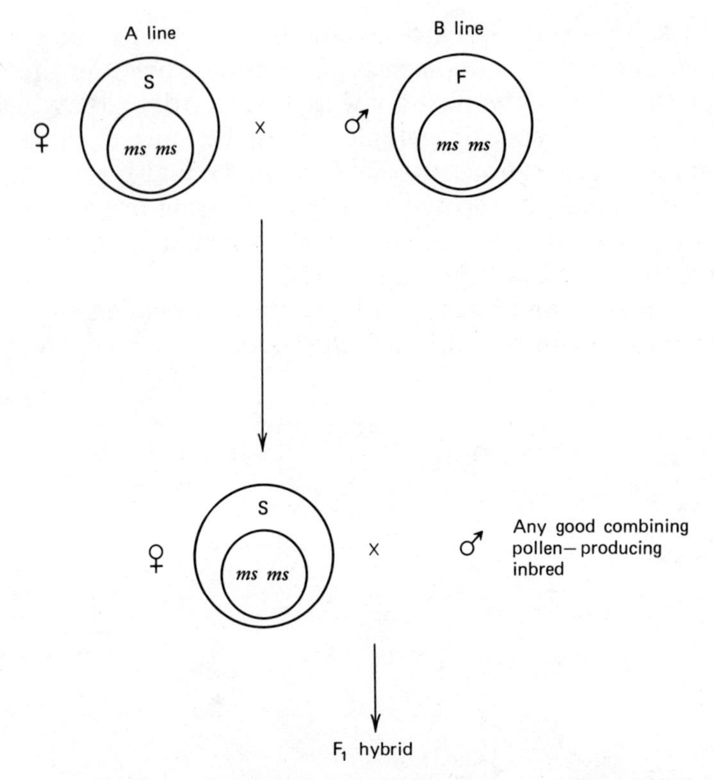

Figure 16.3. Cytoplasmic male sterile propagation and use in a hybrid. S—sterile cytoplasm, F—fertile cytoplasm.

allele (*ms*) in certain cytoplasm (sterile) would result in nonfunctional male flower parts (male sterility), but in a different cytoplasm (fertile) would have no effect. This system contained the potential for hybrid seed production without hand pollination (Fig. 16.3). The male sterile line, commonly called the A line, is propagated by pollination with an identical genotype in a fertile cytoplasm (the B line). The F_1 hybrid is produced by pollinating the male sterile with a fertile cytoplasm pollen-producing inbred that has good combining ability with the sterile line.

The pioneer work in onions opened the door for investigation and use of sterility in other species. For example, cytoplasmic sterility interacting with two recessive alleles was utilized at about the same time to produce hybrid sugar beets. The exact physiological cause of male sterility is unknown, but irregularities in the pollen formation and developmental process result in nonfunctional male gametes. Disruption of the vascular system leading to the anther, absence of starch formation in the pollen, and pollen nuclear division difficulties are observable abnormalities in several species.

Early work in the 1930s demonstrated the possibility of cytoplasmic sterility in corn. However, since corn is a crop dependent primarily on seed production for its economic value, the system required a mechanism to counteract the sterile cytoplasm and produce male fertile plants in the F_1. Without this, male sterile hybrid seed corn would be sold to the producer and the field would have no seed set except for fertile pollen coming in from neighboring areas. Figure 16.4 shows the comparison of a normal pollen-producing corn tassel and one made sterile by a cytoplasmic system. Two sources of cytoplasm, Texas (T) and USDA (S), were found

Figure 16.4. Cytoplasmic male sterility in corn. (*a*) a normal tassel with dehiscing anthers. (*b*) a male sterile tassel devoid of functional anthers.

to have good sterility properties. The male sterile cytoplasm, either T or S, can be counteracted to produce male fertile plants by dominant genes at three loci in the nucleus. The genes, labeled Rf_1, Rf_2, and Rf_3, are independently assorted. A male sterile plant has sterile cytoplasm and recessive alleles at the *rf* loci. Rf_1 and Rf_2 are complementary to each other and must both be present to produce complete male fertility in the T cytoplasm. The presence of one or the other will result in partial fertility. The single dominant gene Rf_3 restores fertility in the S cytoplasm. A higher percentage of corn belt inbred lines was completely sterile in T cytoplasm, and most hybrid production was developed using this source.

The maintenance of the sterility and restorer systems comprises a separate portion of the breeding program. As in onions, the sterile cytoplasm (A line) is maintained by pollinating with the B line that contains the identical genotype, fertile cytoplasm, and no restoration genes. The A line can be used in producing the hybrid without detasseling. A third line, the restorer or R line, contains the restorer genes in either a sterile or fertile cytoplasm and is used as a male to pollinate the A line in hybrid production.

The sterility and restorer systems in no way negate the basic hybrid breeding program and are superimposed on inbred lines after combining ability is determined. Both sterility and restoration are moved from line to line by backcrossing. The sterility is transferred by making crosses with the cytoplasm source as the female. Restorer genes can be backcrossed in the conventional manner, but a periodic testcross to a male sterile line is necessary to insure the presence of restoration alleles.

After conversion of the hybrid seed components to male steriles and restorers, they can be used in a variety of ways. For example, in a single cross the male sterile can be pollinated with an inbred containing homozygous restoration genes. The resulting hybrid will be completely male fertile because of a dominant allele at each locus. Note that the F_2 generation would range from completely sterile to completely fertile because of the segregation of the restoration genes. In a double cross hybrid, male sterility can be introduced into the system in a number of ways. Some hand detasseling may be necessary since it is difficult to produce both of the single cross F_1's with male sterility and yet have the double cross hybrid contain the restorer genes in a heterozygous condition at all loci. A common practice is to produce male sterile hybrids and then mix in a small amount of a genetically identical pollen producer. This is normally adequate for grower field pollination. Regardless of the operational variations, male sterility has reduced hybrid seed costs by eliminating a considerable amount of manual labor.

Because of its monoecious flowering character, corn can be hand emasculated and still produce seed volume on a reasonable economic basis. In most normally self-fertilized crops some type of male sterility system is mandatory if hybrid vigor is to be exploited, since manually made F_1's are cost prohibitive. Sorghum is a good example (3). The discovery of male sterility and restoration led to the development of the hybrid sorghum industry that today occupies over 90 percent of the planted acreage. Another selfed crop currently receiving extensive attention is wheat, where cytoplasmic male sterility and restoration are being utilized in hybrid development. Figure 16.5 shows a production system in hybrid wheat where a male sterile inbred is planted in strips between a pollinator. The F_1 hybrid seed on the male sterile is harvested separately from the pollinator. This system can also be used in the maintenance program with A and B lines interplanted.

Cytoplasmic male sterility has been found in a number of other species including field beans, cotton, flax, tobacco, petunias, red peppers, tomatoes, carrots, rice, and many of the grasses.

Several problems occur in the use of cytoplasmic sterility and restoration. The mechanisms may be environmentally sensitive so that sterility and restoration expression are altered in certain cases (9). For example, a male sterile line may be completely sterile in one situation but have partial self-fertility in a different environment and, likewise, restoration may be complete in one environment but only partial in another. The environmental properties of temperature, photoperiod, and humidity apparently have the ability to interact with the sterility and restorer mechanisms resulting in different levels of penetrance and expressivity. A breakdown in the system can have serious consequences. For example, suppose the components for an F_1 hybrid are being grown in an environ-

Figure 16.5. A hybrid wheat seed production field. The combines are harvesting male sterile strips that have produced F_1 seed by pollination with the male fertile rows. (Courtesy DeKalb AgResearch, Inc.)

ment that promotes a small amount of sterility breakdown, and a low percent of the seed set is produced by selfing rather than crossing with the restorer. If the F_1 seed is then grown in an environment that maximizes sterility expression, the selfed seed will produce sterile plants. In corn, restoration need not be 100 percent to produce a good hybrid crop since an overabundance of pollen is available for fertilization. In self-pollinated crops, on the other hand, there is very little latitude with regard to restoration since some sterile florets or sterile plants in the F_1 can rapidly negate the potential advantage of heterosis.

Minor modifying genes can occur for restoration. In wheat, for example, there appear to be three major restorer genes with a series of minor modifiers necessary to provide complete restoration. Each additional locus added to the system complicates the conversion of inbreds.

A third difficulty arises because of genetic vulnerability. The southern corn leaf blight disease epiphytotic in 1972 came about because T cytoplasm was particularly sensitive to this disease. Since almost all the corn belt hybrids contained T cytoplasm, the majority of the acreage was susceptible. Following the serious crop losses, corn companies discarded the T cytoplasm and reverted to using mechanical or hand detasseling for hybrid production. The search continues for more diverse cytoplasm sources that will circumvent the vulnerability problems.

While male sterile cytoplasm and restoration offer exciting possibilities in hybrid production, they are not without their unique array of associated difficulties. Other production problems exist in converting selfed crops, especially the grasses, to outcrossing. Pollen longevity and movement are usually limited, and florets are not particularly well designed for stigma receptivity to foreign pollination. Genetic variation does occur for most of these characters, however, and progress through selection is continuing.

GENETIC MALE STERILITY

In practically all diploid plant species, recessive alleles regularly occur that produce male sterility without any cytoplasmic interaction. They may also be present in polyploid species but are harder to detect because of some locus duplication. The genetic sterility alleles, designated *ms,* cause male sterility when homozygous recessive, while the heterozygous or homozygous dominant genotypes have normal fertility. This is an extremely simple system genetically and one that has been used to promote cross-pollination in self-pollinated species to maintain and enhance genetic diversity with a minimum amount of labor. By introducing the recessive allele through crossing, and then allowing selfing, a proportion of the plants become male sterile and outcrossing can take place. The allele can easily be removed by selecting homozygous fertile types.

While genetic sterility can be transferred easily from one line to another, its use in hybrid production has one serious problem centering around the maintenance of the male steriles. A male sterile line must be homozygous recessive to be used in making the hybrid. Since the homozygous recessive genotype cannot be maintained alone, it must be pollinated by male fertile individuals carrying at least one dominant allele. If the maintainer is heterozygous, the progeny will be half male fertile, *Ms ms,* and half male sterile *ms ms.* The next step is the sorting of the homozygous recessive from the heterozygotes. This task can be accomplished in several ways. The population can be hand rogued to remove the male fertile plants at flowering time since they can be distinguished by closed florets. Unfortunately, this increases the amount of labor necessary to produce the hybrids and destroys the economic advantage of the male sterile genes. Another possibility is to link the male fertile gene with a locus conditioning chemical susceptibility, and remove fertile types by spraying. An aleurone color gene linkage is also possible to allow electronic color sorting. Another more complicated system is to link the male sterile gene with some type of chromosomal abnormality. A tertiary trisomic chromosome made up of parts from two different chromosomes when added to the $2\underline{n}$ condition results in reduced seed size. With the proper linkage of the male fertile allele on the trisomic chromosome, male sterile seed can be sorted on size from selfed tertiary trisomics. This technique is discussed in more detail in Chapter 18.

Inherent problems in all of these manipulations include the establishment of proper linkages through chromosome breakage and translocation, and difficulty of maintaining linkage relationships without recombination. Patterson (7) has patented a male sterile abnormal chromosome combination in corn. Here the dominant fertility allele is linked with a duplicate-deficient (Dp-Df) chromosome not transmitted through the male. Male sterile seed is harvested from the cross of a male sterile female (*ms/ms*) with the male special stock (Dp-Df *Ms/ms*). The Dp-Df line is propagated by selection within the pollinator. Thus, genetic male sterility can be used in hybrid production but elegant genetic engineering is required to make the system usable.

MALE GAMETOCIDES

Chemical treatment of plants to produce male sterility has been investigated for several years. This would allow the instant development of male sterile inbreds and would eliminate the portions of the hybrid programs concerned with development and maintenance of male sterility and restoration. Frankel and Galun (5) review the use of chemical male gametocides. Ethrel (2-chloroethylphosphonic acid) has been used to induce male sterility in several cereals. The chemical is applied prior to anthesis and results in sterile anther development. The use of gibberellic acid as a male gametocide has been successful in onions. Sodium 2,3 dichloroisobutyrate (DCIB), sodium dichloroacetate, and others have also been used in several crops.

Two problems have been encountered when various chemicals are used. First, the determination of the critical stage of plant development and correct application rate has been difficult. If the chemical is applied incorrectly, plant defects such as height reduction and spike malformation are experienced along with incomplete male sterility. Second, environmental interaction with the chemical is extensive and hard to predict. Research interest in chemical male gametocides remains high, however, because of the potential benefits associated with removing the dependence on cytoplasmic and genetic systems. Its success and value remain to be determined.

INCOMPATIBILITY

The multiple allele incompatibility mechanism offers some possibilities for use in hybrid seed production. Denna (2), using tomatoes as an example, outlines uses of incompatibility in the production of hybrid seed in normally selfed crops. He suggests that incompatibility is more logical than male sterility for use in crops with potential insect pollination because insects are not attracted to male sterile plants with nonfunctional anthers and will thus not pollinate male steriles. In self-incompatibility, however, both the male and the female produce functional anthers and pollen, attracting insect pollinators to carry out the job of crossing. The self-incompatibility mechanism must be identified in related species and then transferred through a backcrossing program to genotypes normally self-compatible because of a dominant fertility allele. The self-incompatible lines would then be planted as females in a field production system and be pollinated by insects, using a homozygous dominant self fertile line as the male. The resulting F_1 would also be self fertile.

Incompatibility has been used to produce F_1 hybrids in cabbage, kale, and brussels sprouts. It has not been used extensively in other crops and problems such as the gametophytic or sporophytic nature of the incompatibility, the existence of corresponding fertility alleles, and the interaction with environment have not been entirely resolved.

SEX EXPRESSION

Dioecism, in which the male and female flower parts are produced on different plants, is a usable mechanism for hybrid production. In spinach, for example, sex expression is controlled primarily by a pair of sex chromosomes designated X and Y. The XX genotype is female and the XY is male. Segregation of the sexes makes hybrid spinach seed production simple by interplanting the two sexes and roguing out the males. The hand roguing adds a labor feature to the system, but allows the production of hybrid seed if the economic return is large enough. In the 1960s about 35 percent of the spinach seed produced in the United States was hybrid using the dioecious system.

In cucumber plants the gynoecious (only female flowers) condition is governed by a dominant gene that can be reversed in action with gibberellic acid (8). To produce a hybrid, a gynoecious line is planted beside an andromonoecious (bisexual flowers and male flowers) line. All seed formed on the gynoecious line are hybrid. The female is propagated by chemical treatment in isolation, resulting in intercrossing between female plants. In the F_1 field a small amount of monoecious pollinator seed is mixed with the hybrid seed to produce pollen for fruit set. Parthenocarpy (fruit development without pollination) is also controlled genetically in cucumbers. This gene is currently being transferred to gynoecious lines so the monoecious pollinator will not have to be added to the F_1 seed.

SUMMARY AND COMMENTS

A number of unique systems, each with its advantages and disadvantages, have been utilized to reduce costs of hybrid seed production. The economic potential of hybrids will stimulate continued research and development on production systems.

I have watched hybrid development in selfed cereals with much interest. Pollination control problems continue to plague the system but some excellent progress has been made. When one considers that we are attempting to change in a few decades a fertilization system that evolved over several thousand years, the progress rate scale may have to be adjusted a bit. Fortunately, the scientific community and its sponsoring organizations will continue to wrestle with hybrid development and associated genetic engineering.

REFERENCES

1. Craig, W. F. 1977. Production of hybrid seed corn. pp. 671–719. In G. F. Sprague (ed.). *Corn and corn improvement.* Am. Soc. Agron., Madison, Wisc.

2. Denna, D. W. 1971. The potential use of self-incompatibility for breeding F_1 hybrids of naturally self-pollinated vegetable crops. *Euphytica* 20:542–548.

3. Doggett, H. 1970. *Sorghum.* Longmans, Green & Co., London.

4. Duvick, D. N. 1966. Influence of morphology and sterility on breeding. pp. 85–124. In K. J. Frey (ed.). *Plant breeding.* Iowa State Univ. Press, Ames.

5. Frankel, R., and E. Galun. 1977. *Pollination mechanisms, reproduction and plant breeding.* Springer-Verlag, Berlin.

6. Harvey, P. H., C. S. Levings, III, and E. A. Wernsman. 1972. The role of extra chromosomal inheritance in plant breeding. *Adv. Agron.* 24:1–27.

7. Patterson, E. 1973. Procedures for use of genic male sterility in production of commercial hybrid maize. U. S. Patent 3,710,511.

8. Robinson, R. W., and T. W. Whitaker. 1974. Cucumis. pp. 145–150. In R. C. King (ed.). *Handbook of genetics.* Vol. 2. Plenum Press, New York.

9. Sage, G. C. M. 1976. The interaction of restoration with environment in wheat. pp. 123–134. In A. Janossy and F. G. H. Lupton (eds.). *Heterosis in plant breeding.* Elsevier Scientific, New York.

17

MUTATION BREEDING

Mutations are the sole source of allelic differences, the raw material for genotype alternatives. They provide nature with inherited variability and are the key to natural selection success.

Point mutations are changes in the chromosomal DNA nucleotide sequence resulting in altered enzyme protein formation. The most common change is the substitution of one nucleotide for another during DNA replication. Since the genetic code is redundant (meaning that several nucleotide triplicates code for the same amino acid), not every substitution results in an altered protein. Another alternative is the gain or loss of a nucleotide, altering the information read from the point of change to the end of the gene. These are more serious changes since they can effect a number of triplicates, and most are likely removed from the population rather quickly by selection.

In this chapter the role of mutations in plant breeding will be considered. While some authors include chromosome rearrangements as mutational events, this discussion deals only with point mutations. Chromosome engineering is covered in Chapter 18.

SOURCES OF MUTATIONS

Natural Mutations

Mutations occur as low frequency natural events. Over a period of time nature evaluates these mutations through testing, and saves the most desirable genetic combinations. Many natural mutations are recessive when compared with the commonly occurring allele (wild type) in the population. The reasons for this are not particularly well understood, but it is possible that long evolutionary periods in most plant species have led to the identification and stabilization of many favorable alleles. Dominant types having a lower fitness value would be eliminated rather rapidly from the population since they appear immediately in the phenotype. Recessive alleles, on the other hand, could be carried in the population and tested in combination with many other allelic forms at different loci before being completely discarded.

Induced Mutations

Following the rediscovery of Mendel's work, interest increased in genetic variability and the potential for altering this variability at will. Gustafsson (3) reports induced mutation research in the early 1900s. Later work with *Drosophilia* demonstrated the scientific nature and feasibility of increasing mutation rates with artificial treatments. We now know that mutational events can easily be produced in the laboratory. This knowledge has led the plant breeder to consider producing mutations valuable for breeding use (4,8).

Induced mutations can be generated with two principal types of treatments: energy and chemical. Included in the energy category are X rays, gamma rays, beta rays, fast and slow neutrons, alpha particles, deuterons, and ultraviolet light. X ray is by far the most commonly used.

The chemical group includes several forms of methanesulfonate, ethyleneimine, diepoxybutane, nitrogen mustards, and ethyleneoxide (8). The ethyl form of methanesulfonate (EMS) has been frequently used. Many other compounds could be included in the list.

Mutation rates vary with mutagen dosage. The higher the dose of mutagen the more frequent the mutations and the greater the associated possibility of undesirable chromosome damage and lethality. A level commonly utilized is the dose at which 50 percent of the treated material is killed. This is called 50 percent lethal dose or LD50. The use of these energy and chemical treatments is a specialized area of genetics, requiring unique knowledge and skill.

The production of induced mutations appears to be nondirectional. This means that a mutagen treatment will alter alleles in a random pattern throughout the genotype. In earlier discussions it was pointed out that some loci mutate more frequently than others, but increasing the rates through induction does not seem to change the relative frequency among loci. The frequency and severity of induced gene change depends on the mutagen dosage, tissue type and age, and physical factors including moisture and temperature. Occasionally, the desired spectrum of genetic variability can be obtained by exposing several generations of plant material to mutagen treatment.

Genetic change can take place anywhere along the chromosome and can include genes for both qualitative and quantitative traits. Gregory (2), studying induced mutations in peanuts, has shown changes in polygenic and monogenic inheritance systems. This increases induced mutation value since additional variability for all traits can be produced.

USE OF MUTATIONS IN SEXUALLY REPRODUCED CROPS

In a breeding program the relative worth of any artificially induced mutation depends on the amount of natural variability available. If the desired allele is in existence, the breeder will likely choose to use the natural form instead of running the risk of unfavorably altering genetic composition and chromosome makeup with mutagens. Artificially induced mutations may be relatively more valuable in self-pollinated than in cross-pollinated species. Selfed plants, in their drive toward homozygosity, eliminate rather rapidly most alleles that do not have a high adaptive value. This in turn reduces genetic variability. In cross-pollinated crops the random mating system maintains heterozygosity within the population structure, resulting in more genetic variability. Thus, the probability of producing desirable mutations and genetic variability by artificial means is theoretically higher in selfed crops. The breeder must compare the genes present with those that could be created by induced mutations.

In sexually reproduced crops seed treatment is the most frequent method of mutation induction. Another common system is the treatment of very young seedlings. In both cases, chimeras result. A chimera is a segment of tissue that has a different genetic makeup than the cells adjacent to it. A mutation in a cell responsible for a plant segment will result in a mutant chimera since two adjacent portions of the tissue have different genotypes. Mutations must occur in meristematic tissue giving rise to the reproductive cells if they are to be passed to the next generation sexually. Graft chimeras occur when the plant tissue is a combination of cells from the stock and the scion, and are not the result of a mutational event.

There are two general methods of using sexually reproduced mutations. They can be selected "as is" without further breeding manipulations through crossing. Here conventional breeding program selection procedures are utilized. The mutagen exposed material called the X_0 generation (also referred to as the M_0 generation) is reproduced by seed with the expectation that favorable variability has been produced. Selection is practiced in the following generations (X_1, X_2, etc.) for the desired allele. If the mutated allele is recessive, it will show up rather rapidly in self-pollinated crops because of the normal drive toward homozygosity. In cross-pollinated crops the selection system may have to go through several generations before the allele is uncovered. In either case, population sizes must be quite large since the mutation frequency, while increased, may not be extremely high. If the mutated allelic form is dominant, it can be identified easily.

The selection system should identify the desirable alleles in a good genetic background. Because of the random nature of induced mutations, the valuable alleles may be in combination with undesirable mutations or with chromosome breakage and abnormalities. Large populations may be necessary to find the correct genotypes.

Mutations can also be incorporated into crossing programs as conventional alleles to obtain the desired genotypes (5). Genetic background problems associated with the random nature of induced mutation systems can be systematically overcome through recombination. Again, the value of the induced mutations is directly proportional to the amount of genetic variability present and the ease with which either type of variability can be used. If a high degree of chromosome abnormality in the form of breakage and rearrangement is associated with the induced mutations, then the breeding value is seriously reduced. The use of mutations in sexually reproduced plants has been largely through selection instead of recombination. This is logical because of the time required to find valuable induced mutations prior to their use in hybridization programs. The latter method is continuing to receive increased emphasis.

USE OF MUTATIONS IN ASEXUALLY PRODUCED CROPS

Breeders and nursery personnel associated with asexually produced crops have appreciated the value of mutants or "sports" for many years (6). This is particularly true in fruit trees and ornamental species. Naturally occurring mutants have been extremely valuable to breeders of tree species because of the time involved in conventional breeding techniques. It has been much easier and quicker to obtain variant plant types from the mutational process than by hybridization and selection.

To be of value in a vegetative propagation system, the mutant must be in meristematic tissue that will reproduce faithfully through cuttings or other vegetative means. Chimeras in epidermal tissue layers will not reproduce in the offspring unless some means of bud stimulation in that tissue layer is accomplished. In many dicotyledenous species the layer of cells immediately under the epidermis is the source of sexual reproductive organ generation. If the mutation occurs in this layer, it will be reproduced through seed mechanisms. The third cell layer is associated with the internal plant mechanisms and root formation. Here, mutations can be propagated through runners, stolons, and bud formation. Thus, the specific location of the mutational event becomes important in vegetative propagation. In many ornamental species segmental chimeras occur from a color gene mutation early in flower formation. The mutation is not propagated vegetatively unless the mutated segments can be isolated and stimulated to produce new plants. Segmental mutations are of interest to ornamental breeders because of their regular occurrence, apparently associated with highly mutable loci.

Mutational events have been particularly important in apple and citrus species (1). In apples, mutations have produced valuable fruit color mutants, and spur or compact mutants resulting in dwarf trees have been sought in all commercial varieties. In citrus the navel orange and the seedless grapefruit, varieties of primary importance, are considered to have arisen by mutation from closely related seedy forms. The "Thompson" pink grapefruit was discovered as a limb sport from a white grapefruit variety. These are only a few of the many examples of valuable material produced by natural mutations in fruit species. Induced mutations are also being utilized in vegetatively propagated crops to shorten the time necessary to breed for desirable characters. In most cases young budwood is the material exposed to the mutational source.

One of the real advantages of mutations in vegetative material is that they need not be passed on through the seed. In fact, in seedless grape and orange varieties mutants must be propagated vegetatively since there is no seed available to produce the next generation. On the other hand, genetic recombination is impossible without sexual alternatives.

PROGRESS WITH MUTATION BREEDING

A comprehensive evaluation of mutation breeding efforts is provided by Sigurbjornsson and Micke (7). Through 1973 they indicate that 98 crop varieties and 47 ornamental varieties were produced by induced mutations. Of the crop varieties, 85 were released from mutational events by selection only, and 13 were developed through hybridization programs. Many valuable morphological and physiological characters have resulted from induced mutations. Mutations are receiving increased attention in breeding programs and can play a significant role in the production of desirable genetic variability.

SUMMARY AND COMMENTS

Allelic form changes can be produced artificially. Advantages include the generation of variability not now available, and the reduction in breeding time through immediate use, especially in vegetatively reproduced plants. Disadvantages include the additional undesirable variability from random mutational events, and increased population sizes necessary to identify mutants in a suitable genotype background. The utilization of mutation breeding is gaining popularity and acceptance and has provided highly valuable variability in many breeding programs. Induced mutations may be a source of variability lost through natural selection or human intervention.

We have not extensively used induced mutations in hexaploid bread wheat partly because of the variability already present, and partly because the compensating effects from different genomes reduces the efficiency of identifying usable mutations. I think mutational breeding will gain momentum and increase in value as we begin to exploit the genetic variability available and look for new sources of alleles that will add to the plant breeding arsenal.

REFERENCES

1. *Advances in fruit breeding.* J. Janick and J. N. Moore (eds.). Purdue Univ. Press, West Lafayette, Ind.

2. Gregory, W. C. 1966. Mutation breeding. pp. 189–218. In K. J. Frey (ed.). *Plant breeding.* Iowa State Univ. Press, Ames.

3. Gustafsson, A. 1969. A study on induced mutations in plants. pp 9–31. In *Induced mutations in plants.* Proc. Int. Atomic Energy Agency, Vienna.

4. *Induced mutations and plant improvement.* 1972. Proc. Int. Atomic Energy Agency, Vienna.

5. *Induced mutations in cross-breeding.* 1976. Proc. Int. Atomic Energy Agency, Vienna.

6. *Induced mutations in vegetatively propagated plants.* 1973. Proc. Int. Atomic Energy Agency, Vienna.

7. Sigurbjornsson, B., and A. Micke. 1974. Philosophy and accomplishments of mutation breeding. pp. 303–343. In *Polyploidy and induced mutations in plant breeding.* Proc. Int. Atomic Energy Agency, Vienna.

8. *The use of induced mutations in plant breeding.* 1965. Food and Agric. Org. and Int. Atomic Energy Agency Tech. Meet. Pergamon Press, New York.

18

CHROMOSOME BREEDING

Constancy in chromosome duplication and numbers is critical to an orderly passage of genetic information from one generation to another. In Chapter 5, however, the possibilities for changes in chromosome number were discussed. Then, in Chapter 17 the concept of gene alteration through artificial mutation induction was introduced. A third type of potential genetic change is the addition, deletion, or transfer of information within and between chromosomes through breakage and rearrangement.

The science and technology of chromosome engineering has developed rapidly over the past few decades. This chapter will be devoted to breeding applications of genome and individual chromosome alterations.

HAPLOIDY

Haploid plants containing only $1\underline{n}$ chromosome number can be identified and reproduced with a low frequency in any plant species. Haploids were originally considered genetic novelties with little application to breeding programs. However, there has been an increased interest recently in the development and use of haploids, for two reasons. First, a haploid individual containing only one of each chromosome has the potential for immediate identification and evaluation of recessive alleles, either naturally occurring or artificially induced, without being masked by dominants on a homologous chromosome. More important in breeding, doubled haploids are completely homozygous diploids produced without the time consuming task of selfing or sib-mating and selection over many generations.

Haploids can be generated by parthenogenesis in the female where the unfertilized gamete or some other haploid female tissue is stimulated to grow mitotically. The more common technique is by androgenesis where the haploid individuals are produced from the stimulation of anthers or pollen grains. Techniques including media culture methods are provided in an extensive 1974 review (3). In haploid propagation from anther culture, care must be taken that the individuals produced are not the result of diploid tissue stimulation.

Haploidy has been produced in a number of crop species including cotton, strawberries, corn, tomatoes, barley, tobacco, potatoes, rice, petunias, geraniums, wheat, and others. The use of the chromosome doubling agent, colchicine, has aided greatly in haploid work. The value of doubled haploids in breeding is illustrated by Park et al. (5) who studied 52 doubled haploid lines in two barley crosses in comparison with the F_6 generation of pedigree and single seed descent derived lines. They found essentially the same population composition for variability of yield, heading date and plant height by all methods, indicating that doubled haploids were a very rapid method of producing usable populations for breeding and selection.

The creation of genetic variability through artificially induced mutations can be accomplished in haploids by exposure to conventional mutagens such as X rays. Pollen treatment and subsequent culture, for example, can result in the instant appearance of any mutational event. The mutants can be evaluated either in haploid or doubled diploid progeny (3).

Haploidy is not without difficulties. First, training and skill are necessary to properly manipulate the techniques of haploid production. Second, chromosome instability, especially with material exposed to mutation induction, can result in low success rates and abnormalities such as aneuploidy. Third, the breeder may have more conventional variability than can be adequately screened and evaluated. In corn there is no need to produce more instant inbreds if the ones already available have not been adequately tested. While these problems are present, they do not negate the consideration of haploidy as a potential plant breeding tool. Collins (1) provides a comprehensive review of anther-derived haploids and their potential role in crop breeding and genetics.

AUTOPOLYPLOIDY

Autopolyploidy occurs occasionally in nature as a mechanism for improving genetic information. For example, alfalfa is an autotetraploid, and commercial bananas and some varieties of apples are autotriploid. Autopolyploidy chromosome movement resulting in an uneven chromosome distribution to the poles presents problems that are sometimes difficult to overcome. Most of the autotriploids are sterile, while the autotetraploids can produce good seed volumes.

In some plant species a reasonable increase in numbers of genomes is accompanied by an increase in cell size and larger organs. Triploid apples, for example, are more vigorous and tend to have larger fruit than the diploids. Since they produce very poor pollen, orchard production requires the presence of two different diploids to pollinate the triploid and each other. In many species, on the other hand, reduced vigor and size is associated with increases in chromosome numbers through autopolyploidy, indicating the delicate balance in most plants for numbers of chromosomes within each cell.

The artificially produced autotetraploid rye variety, Tetrapetkus, was originally produced in Europe and has been grown on an appreciable acreage in the United States and Canada. Autopolyploidy has been utilized as a breeding technique in the development of seedless watermelons (2). Here autotetrapoloid plants are produced by treating diploids with colchicine and are then used as the female for pollination by a normal diploid. The resulting seed is triploid and produces mostly abnormal gametes. Seedless fruit are set when the triploids are pollinated with a diploid. Seedless watermelon have been marketed commercially for some time, especially in Japan. Irregular fruit shape and internal development and specialized seed production requirements are problems in the industry.

Autopolyploidy has been used to produce gigantism in fruits and ornamentals. These are particularly useful where plants can be propagated by vegetative means so that the chromosome abnormalities need not be transmitted through gametes for reproduction.

ALLOPOLYPLOIDY

As discussed in Chapter 7, the combining of genomes from different species has been an important mechanism in evolution and has resulted in the development of many useful plant species. Chromosome doubling of the F_1 following genome hybridization is the normal procedure.

Breeders have been interested in the possibility of combining genetic variability from different species as a breeding tool. A classic example is the cereal grain crop triticale produced by crossing wheat (*Triticum*) and rye (*Secale*) (9). Two types of triticale have been developed. The octaploid, containing six genomes from wheat and two from rye, is generated by crossing bread wheat ($2\underline{n} = 6X = 42$) and diploid rye ($2\underline{n} = 2X = 14$). The F_1 of this cross has three wheat genomes and one rye genome with virtually no pairing among any of the 28 chromosomes. The F_1 is treated with colchicine, chromosome doubling occurs, and the subsequent individuals are $2\underline{n} = 8X = 56$. The second, more common, type is hexaploid produced by crossing durum wheat ($2\underline{n} = 4X = 28$) with diploid rye. Following F_1 doubling, the resulting amphidiploid is $2\underline{n} = 6X = 42$. Hybrids of both types have been in existence for over 100 years. Recently improved types show promise for stress environments and better nutritional properties when compared with their parental counterparts.

A number of difficulties have surfaced with triticale. Abnormal chromosome numbers often occur in the progeny and a high degree of sterility can result. In addition, the seed size and shape has been very poor in many of the progeny. Good selection progress is being made in overcoming the problems, but the development of a new species through allopolyploidy logically takes a great deal of time and patience on the part of the breeder. For this reason it has not been used extensively in most breeding programs.

CHROMOSOME ENGINEERING

Often, breeders are interested in the alteration and improvement of individual chromosomes instead of the more radical approach of restructuring the species with entire genomes. Some alternative methods are considered in this section.

Introgression

The process of introgression—the incorporation of a small amount of genetic material from one species to another—can be accomplished by repeated backcrossing of the interspecific F_1 to one of the parents. The final result may be the inclusion of a whole chromosome either as an addition line in which the resulting plant has an extra pair ($2\underline{n} + 2$), or as a substitution line where one of the chromosomes in the recurrent parent has been eliminated and a chromosome from the donor species has been included ($2\underline{n} - 2 + 2$). Small amounts of information can be incorporated through chromosome breakage and healing. These are translocations and can vary in size from single genes to large parts of whole chromosomes.

Many classic cases of introgression involve the need for disease resistance genes not available in cultivated species. Knott (4) describes the transfer of disease resistance from the grass genera *Agropyron, Aegilops,* and *Secale* to wheat. Following hybridization, useful genes can be transferred in three ways. First, if the donor species is closely related to wheat, introgression through normal chromosome pairing followed by backcrossing can be accomplished. If the species is not closely related, two methods are possible. As discussed in Chapter 5, genes controlling pairing can be manipulated so nonhomologous chromosomes can synapse and exchange segments. The other possibility is to irradiate the F_1 to produce translocations between the donor and recipient chromosomes. Normally, following the cross and translocation, several backcrosses are needed to stabilize the chromosome complement and normalize meiosis.

Sebesta and Wood (8) accomplished the transfer of insect resistance from rye to wheat using triticale as a bridging species. Here the hexaploid amphidiploid triticale is formed and then backcrossed with wheat once or twice. The heads of the resistant plants from the backcrosses were X rayed and used to pollinate several varieties of wheat from which homozygous resistant translocations were selected. This program represents a combination of polyploidy and chromosome engineering through irradiation.

Hybrid Mechanisms

Special cases of chromosome engineering have been developed in hybrid breeding. The utilization of valuable recessive male sterile genes is extremely difficult because of male sterility maintenance problems. Ramage and Wiebe (6) describe a tertiary trisomic system to aid in solving this problem in barley (Fig. 18.1). The required engineering involves the addition of an extra chromosome, made up of two different chromosomes by translocation, to the barley complement. Chromosome abnormalities are generally not passed on through the pollen, but are transmitted by the female. Thus, both normal and abnormal female gametes are generated by a tertiary trisomic plant, but the only functional male gametes are normal. The selfing of a tertiary trisomic results in a combination of tertiary trisomic and normal seed. The trisomic seed is small and shriveled, and can be sorted by mechanical screening. Cross-over frequency is greatly reduced in the area where the two chromosome parts join. By placing the dominant fertility allele near this break and the recessive allele on the normal chromosome, selfed fertile trisomics will produce shriveled fertile trisomic seed and normal sized male sterile seed. Research is currently underway to refine the system in terms of linkage groups and correct trisomic composition for use in hybrid barley production. This same general concept was patented by Patterson for use of genetic male sterility in hybrid corn production.

These systems represent a great deal of scientific effort with many years in their planning and execution. While success is not always assured because of unknowns relative to chromosome movement and recombination, they do provide potential tools to the plant breeder in hybrid systems development.

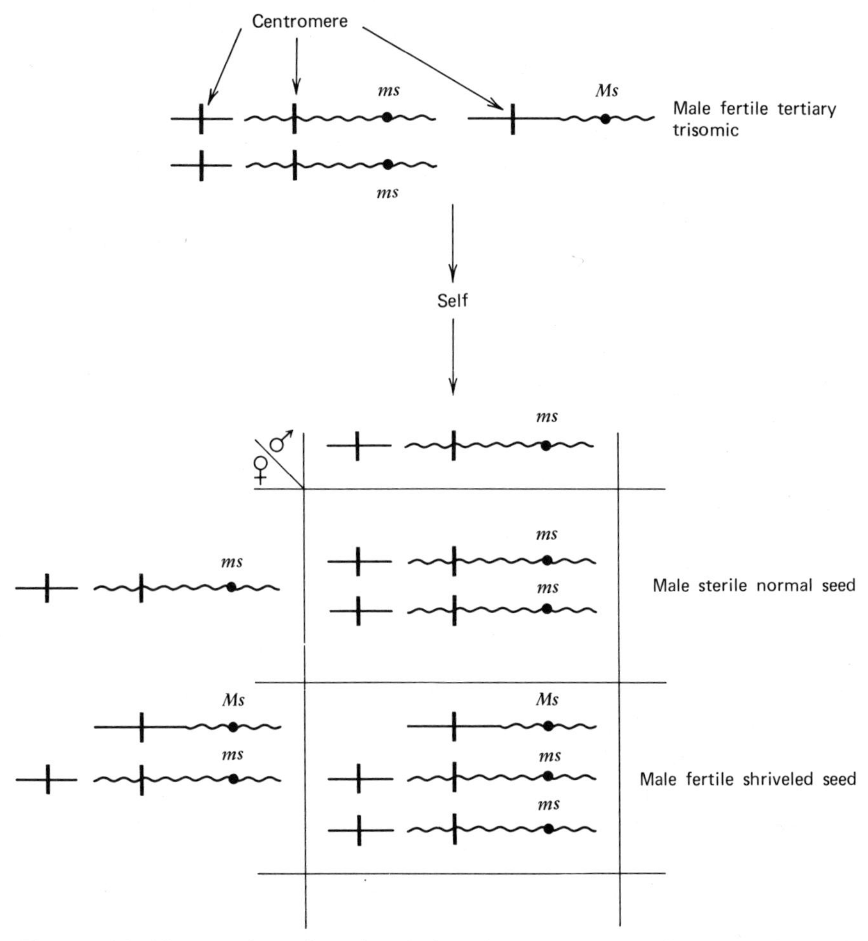

Figure 18.1. The use of a tertiary trisomic in maintenance of genetic male sterile seed. The tertiary trisomic chromosome is made up of parts from two different chromosomes pairs. Assume abnormal chromosome to be passed through female gametes only.

SPECIAL TECHNIQUES FOR WIDE GENETIC TRANSFERS

Recently, several new innovations have been developed to provide the potential for genetic information transfer between species that will not produce viable conventional hybrids. Much of this work has been developed as an offshoot of microbial genetics. At the present time the techniques are in developmental stages, but show promise for practical applications. Those discussed briefly here are reviewed by Scowcroft (7).

Nuclear Fusion and Somatic Hybridization

Crosses between species with widely divergent genotypes are usually not successful. Either the fertilization process is not completed or embryos are nonviable. Protoplasts, the remainder of the cells including the nucleus after the walls have been removed, can be mixed in solution and, with proper treatment, stimulated to fuse. Furthermore, the nuclei, on occasion, will also fuse bringing the genetic information from each parent together in one "cell." Nuclear fusion thus offers the possibility of combining widely divergent genetic information through a process of somatic hybridization, rather than by the conventional sexual process.

Somatic hybridization has been studied in several crosses between tobacco species where hybrids can be produced by conventional breeding techniques. Comparisons, made on a phenotypic basis, between conventional and somatic hybrid progeny prove that nuclear fusion had taken place. The chromosome numbers of the hybrids formed by nuclear fusion have been roughly equivalent to amphidiploids developed by conventional techniques, but in some cases they have been higher than expected, indicating that triple nuclear fusion may have taken place. The hybrid cells were selected from the mixture by growth medium alterations that favored the development of the hybrid over either parent.

While somatic hybridization offers great potential for genetic information transfer, many problems remain to be worked out. One of the most serious is the reproduction rate of different chromosome sets. If the genomes from one species duplicate and divide at a more rapid rate than those from another, the somatic hybrid between the two will have difficulty accomplishing normal division. The selection of hybrid cells from the parents in the culture also requires highly refined methods. On the other hand, introgression following the production of a somatic hybrid may result in the inclusion of a small amount of highly valuable genetic material from one species into another. Both basic and practical aspects of this technique remain to be proven, but it does offer exciting challenges and alternatives in plant breeding.

Recombinant DNA

Very recent advances in modern biochemical genetic techniques have resulted in the transfer of DNA from one species to another. Viruses are commonly used to transform bacterial systems. Virus transfer of genetic information between plant species also offers potential, but the development of this system is still very embryonic.

Gene transfer and manipulation can also be accomplished by using circular double-stranded DNA units called plasmids that are independent of the host bacterial cell chromosomes. Restriction enzymes are also used that have unique capacities to recognize specific short DNA sequences and can cleave the double-stranded DNA within these sequences. Hybrid DNA molecules are formed by the inclusion of foreign DNA into the plasmid system and subsequently into the host chromosome. Many copies of the gene and its product can theoretically be produced. Using this mechanism, it may be possible to transfer genes between widely divergent species. At the present time DNA hybridization is limited to microorganisms and is subject to considerable controversy regarding its potential hazards and value. Should DNA hybridization prove feasible in higher plants, almost unlimited possibilities are available for change and manipulation of genetically controlled characteristics. A commonly cited example is the potential ability to transform nonnitrogen fixation species into nitrogen fixing plants by proper gene transfer. This accomplishment would have far reaching implications regarding the use of commercial fertilizer and food production potential.

SUMMARY AND COMMENTS

Polyploidy and individual chromosome engineering have received increased attention as means of transferring genetic information from one species to another in meiotically stable genotypes. Our knowledge of genetic mechanisms and information systems is continually expanding. Somatic hybridization and recombinant DNA offer the possibilities for combining genetic information in conventionally impossible crosses.

Every breeder should not revamp every program to incorporate some form of DNA wide transfer, since the ultimate value of many of these systems remains to be proven. Also, one of the major problems with practical utilization of current genetic research is that many of us are lacking skills in the basic techniques. However, the horizons are virtually unlimited for newly trained plant scientists.

REFERENCES

1. Collins, G. B. 1975. Use of anther-derived haploids and their derivatives in plant improvement programs. pp. 359-384. In R. H. Burris and C. C. Black (eds.). *CO_2 Metabolism and plant productivity.* Univ. Park Press, Baltimore.

2. Eigsti, O. J., and P. Dustin, Jr. 1955. *Colchicine—in agriculture, medicine, biology and chemistry.* Iowa State College Press, Ames.

3. *Haploids in higher plants, advances and potentials.* 1974. K. J. Kasha (ed.). Univ. of Guelph, Ontario.

4. Knott, D. R. 1971. The transfer of genes for disease resistance from alien species to wheat by induced translocations. pp. 67-77. In *Mutation breeding for disease resistance.* Int. Atomic Energy Agency Proc., Vienna.

5. Park, S. J., E. J. Walsh, E. Reinbergs, L. S. P. Song, and K. J. Kasha. 1976. Field performance of doubled haploid barley lines in comparison with lines developed by the pedigree and single seed descent methods. *Can. Jour. Plant Sci.* 56:467-474.

6. Ramage, R. T., and G. A. Weibe. 1969. Use of chromosome aberrations in producing female parents for hybrids. pp. 655-659. In *Induced mutations in plants.* Int. Atomic Energy Agency Proc., Vienna.

7. Scowcroft, W. R. 1977. Somatic cell genetics and plant improvement. *Adv. Agron.* 29:39-81.

8. Sebesta, E. E., and E. A. Wood, Jr. 1978. Transfer of greenbug resistance from rye to wheat with X rays. *Agron. Abs.* 61-62.

9. *Triticale: First man-made cereal.* 1974. C. C. Tsen (ed.). Am. Assoc. Cereal Chem., St. Paul, Minn.

19

BREEDING WITH TISSUE CULTURE

All plant breeding programs are faced with the two serious problems of generation turnover time and limited population size. Time is an especially difficult problem in species such as trees where several years can elapse between each generation. Even in annual crops, breeders are constantly trying to reduce time and speed up the program by using greenhouse and winter-summer alternate growing sites. Compromises must always be made between the ultimate population size, and the realistic size that can be handled economically in the program. Missing the low frequency highly desirable genotype is a major breeding concern with limited numbers of individuals. Recent advances in tissue or single cell culture offer aid in these problems (4). This chapter briefly discusses the potential of tissue culture in plant breeding.

TECHNIQUES OF TISSUE CULTURE

Tissue culture is basically a system of growing many undifferentiated cells with the capacity to regenerate new plants. The process is normally started with the production of callus (Fig. 19.1*a*). Callus can be generated from many different types of plant tissue including leaves, stems, and roots. To be of value the callus cells must be "totipotent," meaning that they have complete genetic information and the capacity to regenerate a plant with all organs differentiated. The callus is usually generated on agar media containing the proper combination of plant hormones and other chemicals depending on the species.

Figure 19.1. Tissue culture techniques in grasses. (*a*) Undifferentiated callus growth on agar medium with appropriate nutrient supplements. (*b*) Stimulated root growth without shoots. (*c*) Stimulated shoot growth without roots. (*d*) Complete plant regeneration. (Courtesy S. Ladd, Colorado State Univ.)

Following callus tissue generation, the cells can be broken up and transferred to a liquid medium and grown as single cell suspensions. Each species has its own environmental requirements, determined by experimental techniques, to allow normal cell reproduction in the suspension medium. During this stage, tremendous numbers of potential individuals are generated. Nabors (3) has indicated that upwards of 1×10^7 cells are present in 100 ml of tobacco suspension in an actively growing culture. Generation time is a matter of days, rather than months or years.

The final and very critical stage is the regeneration of new plants from individual cells. Either callus cells or single cell suspensions can be used. Regeneration has presented the greatest problem in the technique. Chemical and hormone balance must be exactly correct to stimulate the differentiation of all plant parts from one individual totipotent cell. In many cases either roots (Fig. 19.1*b*) or shoots (Fig. 19.1*c*) can be obtained, but not both. Each plant part requires its own special set of conditions to promote regeneration. Even if complete plants can be reproduced (Fig. 19.1*d*), the techniques in transplanting to a greenhouse or field situation for completion of the life cycle are very critical. According to Scowcroft (5), plants have been completely regenerated in tobacco, sugar beets, wheat, corn, sugar cane, barley, oats, coffee, cassava, potato, and celery. Murashige (2) lists a large number of plant species with potential for regeneration from callus culture.

TISSUE CULTURE USES

Tissue culture may be used in a number of ways in plant breeding and improvement programs (1). First, the large number of individuals in a suspension culture offer the potential for identification of desirable mutants, either natural or induced, in diploid cells. Nabors (3) has demonstrated the development of highly salt tolerant tobacco genotypes selected by gradually increasing the salt levels of the suspension medium. He obtained plants that were able to grow in salt levels approaching sea water. This system gradually eliminates those individuals with unadapted genotypes. Selection occurred for natural mutations existing in the population and advancement was achieved because of the ability to screen large numbers of individuals.

Mutations can also be artificially induced in the population, and selection pressure applied for their isolation and increase. Mutation treatments can be applied to the tissue prior to callus formation, the callus itself, or the cell suspension. Characters that may be selected include tolerance to various chemicals such as herbicides, and disease resistance. Disease resistance could be accomplished by isolating individuals who survive when exposed to compounds produced by the pathogen. It may also be possible to select for such general characteristics as drought tolerance by increasing the osmotic potential of the suspension medium and selecting cells capable of functioning under low water potential. To appear in diploid tissue culture, cell mutations must either be dominant or both alleles at a locus must mutate simultaneously. Anther culture, discussed in Chapter 18, is another form of tissue culture using haploid cells. Here mutations, either dominant or recessive, will appear immediately.

Many unknowns still exist in tissue culture breeding techniques. The stability of chromosomes, both in composition and number, over time in tissue culture has not been completely determined. Mutation rates may be different in tissue culture from those under natural conditions. The induction of mutations still has the problem of being random, and unwanted alleles may be a frequent event. The identification of phenotypic response at the cellular level does not insure that the same response will occur at the whole plant level. The final breeding and selection will still require field evaluation to identify adapted genotypes for natural production conditions. Skirvin (6) reviews the potential of tissue culture in mutation work.

A final use of tissue culture in plant improvement capitalizes on the fact that many copies (clones) of an individual can be produced very efficiently and cheaply. For example, virus-free plants could be multiplied quickly after a disease-free individual is isolated. Also, rare desirable genotypes could be increased rapidly and economically following their identification either by tissue culture or conventional selection procedures. This represents a valuable savings in time and economic resources that could then be put to other uses.

The actual utilization of tissue culture in plant breeding remains to be tested and evaluated. The recent advances made in the technique, however, encourage its consideration as a powerful method of population development.

SUMMARY AND COMMENTS

Tissue culture lends itself to the manipulation of very large numbers of individuals with a minimum of financial investment. The technique involves the generation of callus tissue, cell multiplication, and regeneration of whole plants. It offers potential for the identification and selection of valuable natural or induced mutants and can also aid in multiplication of desirable genotypes. Technical problems remain to be solved in each phase of tissue culture.

There is exciting potential offered by this technique and the most intriguing aspect is the extensive numbers that can be handled in the suspension medium. As a plant breeder I am continually fighting the problem of limited numbers in the field. If breeders can join with tissue culture specialists, the opportunities for achievement are almost unlimited.

REFERENCES

1. Bottino, P. J. 1975. The potential of genetic manipulation in plant cell cultures for plant breeding. *Radiation Bot.* 15:1-16.

2. Murashige, T. 1974. Plant propagation through tissue culture. *Ann. Rev. Plant Physiol.* 25:135-166.

3. Nabors, M. W. 1976. Using spontaneously occurring and induced mutations to obtain agriculturally useful plants. *BioScience* 26:761-768.

4. *Plant cell, tissue, and organ culture.* 1977. J. Reinert and Y. P. S. Bajaj (eds.). Springer-Verlag, New York.

5. Scowcroft, W. R 1977. Somatic cell genetics and plant improvement. *Adv. Agron. 29:39-81.*

6. Skirvin, R. M. 1978. Natural and induced variation in tissue culture. *Euphytica* 27:241-266.

20

RELEASE AND MARKETING

Release and distribution are the culmination of the breeder's effort to put a hybrid or variety in the hands of the producer. Without success in this stage of the program, all prior effort has been wasted. Regardless of the work put into development, the success or failure of a new variety or hybrid rests with the grower who uses experience and judgment to compare the new variety against those already available to the management system. This chapter deals with the introduction of genotypes into the commercial market. Seed production and processing are reviewed by Copeland (2), Gregg et al. (3) and the *1961 Yearbook of Agriculture* (4).

RELEASE AND NAMING

When a breeder has developed a genotype judged to be a candidate for release to the growers, several decisions must be made regarding its fate. Should the genotype be released or discarded? Should it be recommended for particular production areas if it is released? How will the genotype be identified?

These and other questions are usually considered by a naming and release committee within each organization. The breeder rarely, if ever, makes these decisions unilaterally. In public programs the committee structure may include, in addition to the breeder, other scientists, extension personnel, producers and seed growers, utilization industry representatives, and administration. In private companies, the release group may include breeders, other scientists, administration, and sales staff.

The breeder presents information on the proposed selection collected over the period of testing and evaluation. This includes yield performance, quality, disease reactions, maturity, morphological characters, and all other available data. The breeder is obligated to provide a complete summary including deficiencies as well as advantages of the variety. The completeness and accuracy of the data are critical in the release decision. If a variety or hybrid is released with some serious deficiency, the grower can suffer severe economic loss. This is especially dangerous if the variety is successful for the first year or two following release and becomes established on large acreages, and then experiences some serious loss such as an unexpected disease attack or winterkill. Public releases are generally increased and distributed on a smaller scale than those of a commercial company. Thus, public varieties normally experience much more extensive testing by the producer before large acreages are developed. Nevertheless, the release decision based on accurate data is a critical factor that must be carefully considered.

Hybrids and varieties are distributed with particular areas of adaptation in mind. This may be based on maturity characters, disease resistance, response to moisture, or other properties of the genotype. Until the last decade most variety releases included recommendation of the specific area for production. More recently, the genotype is described as accurately and completely as possible with performance data from many geographic locations. The decision is then left to the growers about the usability of the genotype for their management systems. Almost without exception, growers will try a new variety on a limited area for a year or two and conduct their own evaluations.

The collection of data differs considerably between the U. S. and European breeding programs. In the U. S., each individual program, public or private, is responsible for gathering the information used in variety release. The selection may be tested regionally in several states, but the final release decision is based on data summarized by the individual breeding project. In Europe, on the other hand, the common practice is to have government conducted official tests in which candidate selections are tested for several years. The release decision is based, in large part, on successful competition of the selection in official trials for a certain number of years. In some cases the developers must pay for selection testing in the official evaluations. The European system provides more total data, but also involves additional time in the final testing stages.

If the release committee approves a new genotype, the next step is to provide a name. The naming method differs from program to program. Towns, regions, landmarks, or cultural history situations are often used. In some cases varieties are named for former breeders or other scientists who have been dedicated to agricultural research. Yet another system assigns numbers to varieties so that an individual state or program has an identification name followed by a succession of numbers for new releases. Normally names are chosen so that they are easy to pronounce and spell. In the U. S., agronomic crop names are commonly cleared with the USDA Plant Germplasm Center, Beltsville, Md., to determine if they have been used in the past since repetition adds confusion to the system. Several names are submitted to the release committee and one is chosen for assignment to the variety.

Public agencies will sometimes release germplasm without official naming, for use by public and private breeding programs. This is done to aid in distribution and utilization of valuable genes for disease and insect resistance and other characters. The originating program provides the scientific input necessary to develop the germplasm and then leaves the breeding responsibility to those organizations that are more directly concerned with variety or hybrid release. Companies will often contribute financially to some type of foundation or seed association, which in turn aids public programs in financing their research.

In the U. S., agronomic varieties and germplasm are registered with the Crop Science Society of America, with descriptions appearing in the *Crop Science* journal. This becomes a permanent record of germplasm and varieties and can be used for future reference in identification of valuable breeding material and germplasm ownership.

PROPAGATION

Concurrent with naming and release, the plant material is increased to go to the producer. This is a very critical phase since it involves the reproduction of germplasm representing the variety. The subject can be divided into the two categories of seed crops and vegetatively reproduced crops.

Seed Crops

A variety is sold as a describable and identifiable population. The genetic composition of any plant population can change over time for several reasons. First, natural mutations can occur with low frequency and have the potential to increase over several generations, altering the genetic composition of the crop. Second, some levels of outcrossing may occur, even in the self-pollinated species, resulting in genetic recombination and phenotypes not representative of the original variety. Third, growers and seed producers can experience mechanical mixtures as they produce their seed from generation to generation. This may come from harvesting and planting equipment, from volunteer plants, faulty cleaning equipment, and many other causes.

Because the genetic composition of a variety can change, the seed industry has developed a method to insure that seed is accurately named and described in purity and viability. The seed certification system has evolved first on an individual state basis and more recently on a national level.

While the specific requirements vary from crop to crop, the general principle of certification involves seed classes that are divided into four categories (Fig. 20.1). Breeders seed comes from a small increase that the breeder feels is a true genetic representation of the variety. The production of breeders seed resulting from several hundred selected head or plant rows in small grains, for example, may be no more than 400 to 500 kilograms. Breeders seed is planted in ground not subject to volunteer plants of the same crop, and kept free of weeds. The field is rogued several times during the season for off-types. The harvest from this crop is bulked and is called foundation seed. In both public and private organizations foundation seed is produced by a specific project.

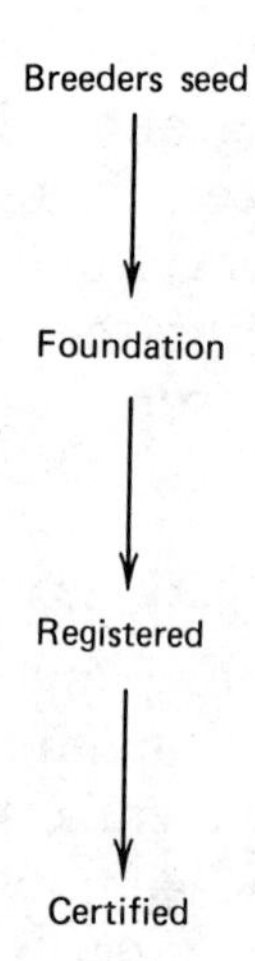

Figure 20.1. Certified seed classes.

Foundation seed is normally turned over to a small group of highly experienced growers who plant it to produce registered seed. Registered seed is planted to provide certified seed that is the stage at which adequate amounts are available to satisfy the needs of most producers. Note that four years are involved from the time of variety release to the build-up of commercial stocks. This is time added to the end of the breeding program before a variety is grown on a large scale.

Strict standards of genetic purity, freedom from weeds and other crops, and seed viability are employed in each of the certified seed classes. Figure 20.2 shows special precautions that must be taken to insure seed purity in cross-pollinated crops. In this case mechanical isolation through caging has been used to protect against foreign pollen. In larger plantings spatial isolation is the usual practice. The seed is followed from purchase through production and processing to the sale for the next generation. Certification labels (Fig. 20.3) showing variety name, lot number and seedlot characteristics are issued when the seed has successfully passed all inspections and tests. The tags must accompany the seed at the time of sale.

Until recently, seed was handled in bags. With modern machinery and larger seed requirements by each producers, bulk han-

Figure 20.2. Legume seed plots caged to prevent outcrossing. A hive of honey bees is placed in each cage to cross-pollinate the plants and produce seed. (Courtesy C. M. Rinckner, USDA-SEA.)

dling is now possible. Labels are still issued, but are identified with large lots instead of individual units.

The certified seed system operates on a limited generation basis—which means that only a lower level of seed can be produced. For example, registered seed can produce only certified, and certified seed cannot be recertified. Thus, only four generations can be reproduced before returning to the original breeders seed to maintain the genetic integrity of the variety. Growers have the option of reproducing their own seed, especially in self-pollinated crops. If they elect to do this, however, once the certified seed class has been reached the generations following are not eligible for certification. Many growers will allocate a small amount of their production to registered or certified seed each year and use this for planting their commercial operations. This allows them to reproduce their own seed for one or two generations while maintaining a high level of genetic purity.

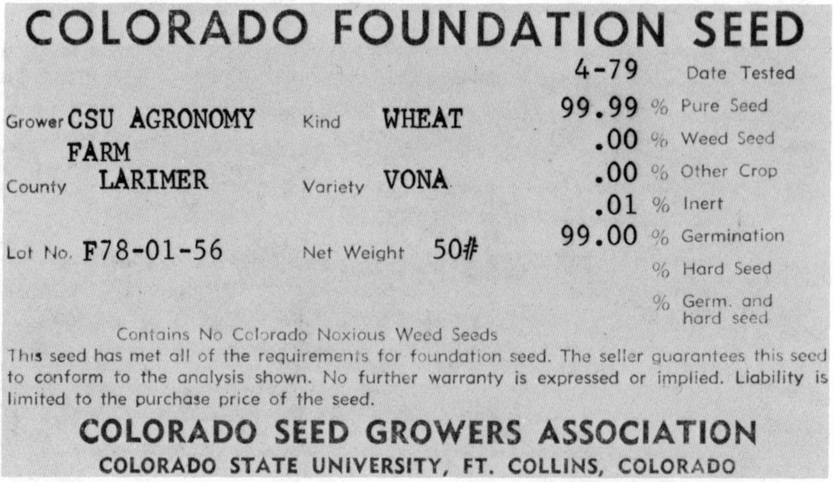

COLORADO FOUNDATION SEED

				4-79	Date Tested
Grower	CSU AGRONOMY FARM	Kind	WHEAT	99.99	% Pure Seed
				.00	% Weed Seed
County	LARIMER	Variety	VONA	.00	% Other Crop
				.01	% Inert
Lot No.	F78-01-56	Net Weight	50#	99.00	% Germination
					% Hard Seed
					% Germ. and hard seed

Contains No Colorado Noxious Weed Seeds

This seed has met all of the requirements for foundation seed. The seller guarantees this seed to conform to the analysis shown. No further warranty is expressed or implied. Liability is limited to the purchase price of the seed.

COLORADO SEED GROWERS ASSOCIATION

COLORADO STATE UNIVERSITY, FT. COLLINS, COLORADO

Figure 20.3. An identification tag used in the certified seed industry. The tag lists the crop, grower, variety, lot number, and the properties of the seed. Issuance of the tag indicates that the seed has passed all certification requirements.

The requirements involved in seed production are very carefully laid out by the Association of Official Seed Certification Agencies (AOSCA). The standards, updated annually, are listed in the AOSCA Certification Handbook Publication No. 23, which is available from all official seed certification agencies and university libraries. Growers must be aware of each requirement from seed purchase, through production, testing, and sale. Failure to comply with the requirements results in a potentially profitable seed crop being rejected for certification and placed on the open market at reduced prices. Rules for each crop are available through state seed certifying agencies and should be studied carefully before the grower enters the business.

The F_1 hybrid industry can also participate in the seed certification system. Normally hybrid companies maintain their own stocks of inbred lines that are eligible for certification through a state agency. Since the companies control the inbred stock and the subsequent sale of the hybrids, many of them practice their own form of seed certification that insures the uniform quality and high level of performance necessary for repeat customer sales.

In the U. S., seed production of many crops occurs in areas other than those where they will be commercially grown. This is true for the legumes and grasses where seed is produced in one area, certified, and transported to another area for use in commercial production. The western U. S. has expanded in the seed business primarily because of desirable environments for pollination, and reduction in storage and processing problems with reduced humidity levels.

Special problems exist in seed production. For example, in some areas a common practice has been to burn grass seed fields following harvest to control diseases. With public demands for environmental protection, legislation has been passed restricting burning in the grass seed producing areas. While the problem has not been entirely resolved, it could result in the possible movement of seed production to some other geographic area that is either disease free or is less strict on production factors.

Vegetatively Reproduced Crops

Vegetative propagation includes cuttings, tuber production, grafting, runners, and other methods common to many of the vegetables, ornamentals, and fruit trees. General requirements include faithful reproduction from one vegetative stage to the next and freedom from disease.

The first of these is normally no problem unless mutations occur that result in some genetic change. The nursery or company reproducing the material must constantly be aware of mutants and remove off-types. If the mutation is desirable, it can be selected and reproduced into a new variety.

The disease problem is more serious. Many diseases transmitted through vegetative material are caused by viruses. The maintenance of disease free stocks plays an important economic role in the company and often requires professional pathology personnel to carry out this portion of the program.

Regulatory agencies and programs in vegetative material are usually found at the state level. California, for instance, has developed extensive programs in this area because of the large concentration of vegetatively reproduced crops. In potatoes, the material is carried through normal certification channels with major emphasis being on freedom from disease.

Nursery stock standards have been developed by the American Association of Nurserymen (1). These standards are descriptors of plant development and include size, shape, and general health at the time of sale. They are used in roses, for example, to describe the different grades on the market, but do not involve genetic purity and freedom from disease.

MARKETING

The vendor varies according to the development program and the type of crop. Until plant variety protection, the primary foundation seed source in self-pollinated species has been the state experiment stations. This is because private industry has not been interested in the development and sale of plant material that can be easily reproduced by growers for several generations. The standard system is the production of foundation seed by the originating organization, followed by grower production of registered and certified seed. The seed growers often have their own organizations but work very closely with the public agencies.

In the case of cross-pollinated crops and those where the seed market is very limited, public agencies may develop varieties and then enter into an exclusive release agreement with private seed companies. A private company with exclusive control of the germplasm is then willing to devote the necessary financial resources to marketing and promotion, resulting in seed availability to producers. The public agency may enter into a royalty agreement with the company so that at least part of the development costs are recovered through seed sales. Here the public breeding program and the private seed company work together to make the best genotypes available to producers.

In hybrid seed crops the companies who control the inbreds are exclusively responsible for seed sales. Occasionally, public agencies will develop and maintain inbred lines that are then sold to private companies for their incorporation into specific hybrids. This is a common practice in corn and allows small companies to utilize public research while reducing their own development inputs. However, they may be limited in their number of marketable hybrids.

In the seed business, success depends on grower satisfaction with the product purchased. It is highly self-regulatory and is based on repeat sales to satisfied customers.

ORGANIZATIONS IN PLANT PROPAGATION

Over the years several important organizations have evolved for seed production and regulation. The Association of Official Seed Certifying Agencies (AOSCA) represents the seed certifying agencies in the United States and Canada. Formerly known as the International Crop Improvement Association (ICIA), AOSCA has the primary function of establishing and publishing the minimum standards for certification of all crops reproduced from seed. The standards, accepted by all state member agencies of AOSCA, represent minimums that must be met if seed is moved through interstate commerce channels. The standards adopted by AOSCA are also those contained in the Federal Seed Law.

The Organization for Economic Cooperation and Development (OECD) encompasses several European countries, Japan, and North America, and establishes a mechanism for the varietal certification of seeds for the international market.

Each state has a certifying agency created by the state government. They may be independent organizations or associated with experiment stations and land grant universities. The Cooperative Extension Service oftens works very closely with the state certification programs. Each agency has minimum standards for certified crops. While these may differ from those established by AOSCA, the seed that moves between states must meet AOSCA minimums. Growers interested in any of the certified seed classes should consult their state agencies for all minimum standard requirements.

The National Foundation Seed project is a cooperative effort of USDA, the State Experiment Stations, the seed industry, and certification agencies to aid in the development of forage seed production. This organization has been helpful in cases where the breeding programs may not be able to produce seed because of environmental limitations. Indiana, for example, has had very productive alfalfa breeding programs, but is environmentally limited in economic seed production. Seed is produced in western areas and returned to the Indiana agricultural system following certification.

The Association of Official Seed Analysts (AOSA) has been instrumental in the development of the seed certification industry. During the certification process official laboratory analyses must be conducted to determine the purity, soundness, germination percent, and viability of the candidate seedlot. AOSA has established uniform standards for each test category in all crop species. Sophisticated techniques such as staining procedures for viability determination in hard seeds have been developed. Official laboratories may be either independent or associated with a land grant system or experiment station.

LEGISLATION IN PLANT PROPAGATION

A number of significant legislative acts affecting the seed and plant propagation industry have been enacted. In 1939 the Federal Seed Act involving truth in labeling was passed to control seed imported and moved through interstate commerce. Labeling requirements are those currently expressed on standard seed certification tags or labels. The standards contained within the Federal Seed Act are also those adopted by AOSCA. An individual state may have seed standards lower than those of the Federal Seed Act, but the seed not meeting the federal regulations may not be moved across state boundaries. Failure to comply with the regulations can result in "stop sale" orders and possible fine and imprisonment. Normally, violations of the Federal Seed Act are identified by state organizations and then reported to the Federal Seed system.

Asexually produced crops have been covered legally by plant patent laws since 1930. As with all other patent applications, the originator must prove uniqueness and originality and must conform to general patent laws. Plants such as roses have a long history of patenting.

The most recent legislation having significant effect in the seed industry is the Plant Variety Protection Act of 1970. This act was stimulated by the need for protection of the originator of self-pollinated varieties. Prior to this time, breeders of self-pollinated crops had no legal protection against individuals reproducing and selling the variety. The Plant Variety Protection Act allows the breeder to apply for and receive legal protection on any self-pollinated variety providing uniqueness and ownership can be proven. Following the granting of protection, the breeder can maintain exclusive rights and obtain royalty payments on seed sales. The Plant Variety Protection Act allows the transfer of ownership from organization to organization, providing appropriate informational procedures are followed. A section of the Act labeled Title V, if invoked by the breeder, makes it a federal offense to sell seed that has not gone through the certification system. This provision has two important implications. First, it allows the breeder a mechanism to follow seedlots from sale to sale and, second, it encourages the use of certification to produce high quality seedlots. Farmers receive a special exemption granting them the right to produce seed of a productive variety for their own use. Protected varieties may also be used for breeding purposes without violation of the Act. Plant Variety Protection has encouraged the entry of commercial companies into the development and sale of self-pollinated crops.

Canadian seed laws are also designed to maintain seed standards in their production system. The Canadian Department of Agriculture is responsible for seed inspection work and law enforcement. Canada uses a variety licensing system to prevent deception by sale of seed under modified or false variety names. Varieties must be tested and found desirable for Canadian agriculture before licensing is issued.

European seed laws are much more restrictive than those of the United States. Extensive government testing, in some cases up to five years, must be completed before the variety can be sold. Also, the European system has very strict morphological description requirements.

SUMMARY AND COMMENTS

The seed laws and certification systems provide mechanisms whereby the efforts of the plant breeder can be extended through seed and plant propagation to the producer. With the establishment of standards in the seed industry, improved products are now being placed on the market.

The seed industry is highly self-regulated. Illicit seed businesses continue to make the agricultural producer aware of the validity of reputable organizations.

The most satisfying aspect of a plant breeder's job is seeing a variety or hybrid go into commercial production, yet the release of a variety does not always insure its acceptance. I have had the pleasure of developing and releasing several successful winter wheat varieties, but have also produced some with poor grower acceptance because of one fault or another. An intriguing aspect of plant breeding is the inability to completely predict the acceptance of every genetic combination.

Plant breeding will continue to serve humanity by providing new useful genotypes. It is a profession filled with variety and unpredictability, but always with the promise of satisfying achievements.

REFERENCES

1. American Standard for Nursery Stock. 1973. Am. Assoc. of Nurserymen, Wash., D. C.

2. Copeland, L. O. 1976. *Principles of seed science and technology.* Burgess, Minneapolis.

3. Gregg, B. R., A. G. Law, S. S. Virdi, and J. S. Bales. 1970. *Seed processing.* Miss. State Univ., National Seeds Corp., and U. S. AID, Washington, D. C.

4. *Seeds.* 1961. *USDA Yearbook of Agriculture.* Washington, D. C.

SCIENTIFIC NAME GLOSSARY

COMMON NAME	SCIENTIFIC NAME
alfalfa	*Medicago sativa*
almond	*Prunus amygdalus*
apple	*Prunus malus*
apricot	*Prunus armeniaca*
asparagus	*Asparagus officinalis*
avocado	*Persea americana*
barley	*Hordeum vulgare*
banana	*Musa sapientum*
bean	*Phaseolus vulgaris*
beet (sugar)	*Beta vulgaris*
begonia	*Begonia spp.*
birds-foot trefoil	*Lotus corniculata*
blackberry	*Eubatus spp.*
blueberry	*Vaccinium spp.*
broccoli	*Brassica oleracea*
bromegrass	*Bromus inermis*
brussels sprouts	*Brassica oleracea*
cabbage	*Brassica oleracea*
carrot	*Daucus carota*
cassava	*Manihot esculenta*
cauliflower	*Brassica oleracea*
celery	*Apium graveolens dulce*

COMMON NAME	SCIENTIFIC NAME
cherry	*Prunus spp.*
chickpea	*Cicer arietinum*
cicer milkvetch	*Astragalus cicer*
clover (alsike)	*Trifolium hybridum*
clover (ladino)	*Trifolium repens*
clover (red)	*Trifolium pratense*
clover (sweet)	*Melilotus officinalis*
corn (maize)	*Zea mays*
cotton (asiatic)	*Gossypium herbaceum*
cotton (upland)	*Gossypium hirsutum*
cress	*Lepidium sativum*
cucumber	*Cucumis sativus*
date palm	*Phoenix dactylifera*
fig	*Ficus carica*
flax	*Linum usitatissimum*
four-o'clock	*Mirabilis jalapa*
geranium	*Pelargonium spp.*
gourd	*Curcubita spp.*
grape	*Vitis uinifera*
grapefruit	*Citrus paradisi*
hemp	*Cannabis sativa*
hops	*Humulus lupulus*
jojoba	*Simmondsia chinensis*
kale	*Brassica oleracea*
lettuce	*Lactuca sativa*
maize (corn)	*Zea mays*

COMMON NAME	SCIENTIFIC NAME
marigold	*Calendula officinalis*
millet (pearl)	*Pennisetum americanum*
nectarine	*Prunus persica nectarina*
oat (cultivated)	*Avena sativa*
oat (wild)	*Avena fatua*
okra	*Abelmoschus esculentus*
olive	*Olea europaea*
onion	*Allium cepa*
orange (sweet)	*Citrus sinensis*
pea	*Pisum sativum*
peach	*Prunus persica*
pear	*Pyrus communis*
peanut	*Arachis hypogaea*
pecan	*Carya illinoensis*
pepper (red)	*Capsicum annuum*
petunia	*Petunia spp.*
plum	*Prunus domestica*
potato	*Solanum tuberosum*
pumpkin	*Cucurbita maxima*
radish	*Raphanus sativus*
raspberry (black)	*Rubus spp.*
raspberry (red)	*Rubus idaeus*
reed canarygrass	*Phalaris arundinacea*
rice	*Oryza sativa*
rose	*Rosa spp.*

COMMON NAME	SCIENTIFIC NAME
rye	*Secale cereale*
ryegrasses	*Lolium spp.*
snapdragon	*Antirrhinum majus*
sorghum	*Sorghum bicolor*
soybean	*Glycine max*
spinach	*Spinacia oleracea*
strawberry	*Fragaria spp.*
sudangrass	*Sorghum sudanense*
sugar cane	*Saccharum officinarum*
sunflower	*Helianthus annuus*
tobacco	*Nicotiana tabacum*
tomato	*Lycopersicon esculentum*
triticale	*Triticale spp.*
vetch	*Vicia spp.*
walnut	*Juglans regia*
watermelon	*Colocynthis citrullus*
wheat (bread)	*Triticum aestivum*
wheat (durum)	*Triticum durum*
wheatgrass (western)	*Agropyron smithii*
wild rice	*Zizania aquatica*

GLOSSARY OF TERMS

"A" line an inbred male sterile line used in hybrid production.

Alleles variations of a gene at one locus.

Amphidiploid a diploid arising from a combination of different genomes.

Androgenesis production of individuals from anthers.

Andromonoecious having only male flowers.

Aneuploidy chromosome composition involving number variations other than complete genomes.

Anthesis the blooming of a flower.

Apomixis production of seed without the union of male and female gametes.

Asexual reproduction new individuals produced by mitotic cell division.

"B" line a male fertile line for maintaining male sterility in a hybrid breeding program.

Backcross crossing progeny with a parent.

Bivalent two homologous chromosomes paired during meiosis.

Breeders seed the initial collection of genotypes representing a new variety.

Bulk a group of unseparated individuals.

Callus actively growing undifferentiated tissue.

Centromere point of attachment for chromosome movement during cell division.

Character trait or form, for example, plant height or flower color.

Chasogomy flowers open after pollination and fertilization.

Chiasma points of very close association between chromosomes of a pair during early meiosis (plural-chiasmata).

Chimera a tissue segment with different genetic makeup than the adjacent cells.

Chromatid longitudinal half chromosome.

Chromosome structure of inheritance contained within the nucleus.

Classification placing individuals into a category or group based on phenotype.

Cleistogamy flower fails to open.

Clone an exact genetic duplicate.

Closed mating system no outside individuals are allowed to introduce gametes into the population. Progeny are produced by known parental combinations.

Combining ability potential to produce a high proportion of desirable individuals.

Composite cross bulking early generation segregates from several populations.

Correlation statistical relationship between two sets of values.

Coupling both of two loci are either homozygous dominant or homozygous recessive in a single individual.

Cross combining gametes from two parents.

Cross-fertilization zygotes produced by gametes from different individuals.

Crossover the exchange of chromatid segments between chromosomes of a pair.

Crossover units the distance in crossover probability between two loci.

Dehisce to release pollen.

Diallel mating each individual is mated with every other individual.

Dihybrid a parental combination that will result in a 9:3:3:1 progeny ratio or some variation thereof. Two different alleles are present at each of two independently assorted loci.

Dioecy male (staminate) and female (pistillate) plants.

Diploid a $2\underline{n}$ individual with normal chromosome pairing.

DNA deoxyribonucleic acid made up of four specific organic bases, sugar, and phosphorus. Contains the genetic code in the cell nucleus.

Dominant a character variation that is completely expressed in the F_1.

Double fertilization the simultaneous process of zygote and endosperm initiation.

Duplex an autopolyploid in which the dominant allele at a locus is present twice.

Ear-to-row the individuals in a single row have been obtained from one ear.

Emasculation removing male flower parts, or rendering them nonfunctional.

Embryo the developing zygote.

Environment the surroundings of an individual.

Epistasis interaction between loci in the expression of a character.

Euploidy chromosome composition involving complete genomes.

Experimental error unexplained variation in an experiment.

Expressivity the degree or intensity of phenotypic expression by a genotype in a single environment.

Family a group of progeny originating from a single source.

Fertilization the combination of male and female gametes to produce a zygote.

F_n the system of identifying generations following a cross. F_1 is the first, F_2 the second, and so on. "F" is the abbreviation for "filial," which means "of or pertaining to a son or daughter."

Gamete a reproductive cell, such as an egg or pollen grain.

Gene a unit of inheritance partially or completely responsible for the genetic control of a character. A specific sequence of bases in a DNA segment.

Gene action the way in which a gene or genes controls character expression.

Genetic code the organic base triplicate combinations contained in DNA that result in specific enzyme formation.

Genetic gain the improvement of a population through inherited variation.

Genetic vulnerability the potential for extensive crop losses because of genetic uniformity.

Genome a basic set or group of chromosomes.

Genotype the genetic makeup of an individual.

Germplasm hereditary material.

Germplasm base the amount of genetic variability present.

Gynoecious having only female flowers.

Head row a single row representing the progeny from one spike.

Heritability the proportion of total variation under genetic control.

Heterogenous a population of different genotypes.

Heterosis vigor in an F_1 hybrid over either parent.

Heterozygous different alleles at a locus.

Homogenous a population of identical genotypes.

Homoelogous chromosomes those that have some similar gene sequences (partial homology).

Homologous chromosomes those that have identical linear gene sequences.

Homozygous identical alleles at a locus.

Hybrid the progeny, generally the F_1 generation, from any parent combination for potential commercial marketing.

Hybridization the act of crossing two parents.

Ideotype ideal plant type.

Inbreeding the mating of closely related individuals.

Inbreeding depression loss of vigor associated with self or sib matings in naturally cross-fertilized species.

Incompatibility failure to accomplish fertilization between certain male and female gametes.

Independent assortment the segregation pattern for one character is independent from that of another character.

Induced mutation a mutation resulting from a specific treatment.

Introgression a small amount of genetic information transferred from one species or genus to another.

Isogenic lines sets of material differing genotypically by only one allele.

Landrace a diverse plant population developed in a specific geographic location by human selection.

LD_{50} the amount of treatment resulting in 50 percent death (lethal dose) of treated individuals.

Linkage two or more loci located on the same chromosome.

Linkage map a map constructed from genetic data identifying the relative location of loci on a chromosome.

Locus the particular place that a gene occupies on a chromosome (plural—loci).

Male sterility condition in which viable male gametes are not produced.

Mass selection the identification of desirable plants to be bulked for the next generation.

Megaspore haploid cell leading to female gamete.

Meiosis cell division resulting in a chromosome reduction to one-half the somatic number.

Microspore haploid cell leading to male gamete.

Mitosis cell division resulting in a duplication of chromosome number.

Monoecious male and female flowers on different parts of the same plant.

Monogenic inheritance one locus controlling one trait.

Monohybrid a parental combination that will result in a 3:1 progeny ratio or some variation thereof. Two different alleles are present at one locus.

Multiline variety a variety composed of a mixture of isogenic lines.

Multiple alleles several allelic forms for one locus.

Mutation a change in the genetic code within a gene.

n formula notation used in mathematical calculations.

n̲ the gamete chromosome number. 2n̲ is the zygote chromosome number.

Nicking either the physical matching of male and female plants for crossing, or the production of desirable progeny from a cross.

Nondisjunction chromosomes of a pair fail to separate during meiosis.

Nonrecurrent parent the nonrepeating parent in a backcross program.

Nucellar embryony seeds resulting from the nucellus rather than the zygote.

Nucleolus circular body visable during early cell division, involved in chromosome organization and cell division.

Nulliplex an autopolyploid in which the dominant allele at a locus is not present.

Parent any individual used in a cross.

Parthenocarpy fruit development without pollination.

Pedigree a system of ancestral lineage.

Penetrance the frequency with which a genotype will exhibit a specific phenotype in a single environment.

Perfect flower male and female parts in the same flower.

Phenotype the appearance or character expression in an individual.

plant row a row representing the random progeny from a single plant.

Plasmid a circular double-stranded DNA unit.

Pleiotropy one gene affecting more than one trait.

Pole outer point of a dividing cell toward which the chromosomes move in later division stages.

Pollination pollen arriving on the stigma.

Polycross allowing selected individuals to mate at random.

Polygenic inheritance several loci controlling one trait.

Polyploid an individual containing more than two genomes.

Progeny offspring in any generation.

Progeny testing measuring progeny to evaluate parents.

Protandry maturation of the stigma prior to the pollen.

Protogyny maturation of the pollen prior to the stigma.

Pure line true breeding genotype.

Qualitative inheritance phenotypic classification of progeny results in a few well defined nonoverlapping classes.

Quantitative inheritance phenotypic classification of progeny result in many poorly defined overlapping classes.

Quartet four cells at the completion of meiosis.

"R" line an inbred line with male fertility restoration to be used in hybrid production.

Random deviation uncontrolled, nondirected variation occurring by chance alone.

Recessive a character variation that completely disappears in the F_1.

Reciprocal crosses crosses between two parents in which each parent is used as the male in one cross and the female in another cross.

Recombination the production of new combinations of different alleles within and among loci.

Recurrent parent the repeating parent in a backcross program.

Recurrent selection identification of superior individuals and their subsequent mating.

Repulsion one locus is homozygous dominant and another homozygous recessive in a single individual.

RNA ribonucleic acid made up of four specific organic bases, sugar and phosphorus. Responsible for information transmission and translation from DNA to enzymes.

Sampling error data errors from a biased sample within a population.

Segregation when a group of progeny from a cross show differences in character expression because of genetic variability between parents.

Selection identifying desired individuals.

Selective value the relative advantage of an allele or genotype in a particular environment.

Self-fertile zygotes can be produced from male and female gametes of the same individual.

Sexual reproduction progeny produced by the union of male and female gametes.

Sib mating brother-sister mating.

Simplex an autopolyploid in which the dominant allele at a locus is present once.

Single seed descent the advancement of generations by one seed per individual.

Somatic cells all nonreproductive cells.

Spindle fiber structure associated with chromosome movement toward poles.

Sport a mutant genotype.

Sterility inability to produce a viable zygote.

Synapsis chromosome pairing.

Synthetic a variety which originates as a known mixture of cross-fertilizing genotypes.

Testcross (hybrids) the cross of a line to be evaluated with an inbred line tester.

Testcross (linkage) a cross, usually of an F_1, to a homozygous recessive individual.

Tester an individual or line used to evaluate or identify the genotype of another.

Tetrad the four chromatids of two synapsed chromosomes in meiosis.

Topcross the cross of a line to be evaluated with an open-pollinated tester.

Totipotent complete genetic information with the capacity to regenerate whole individuals.

Trait character or form, for example, plant height or flower color.

Transgressive segregation progeny occurring outside the phenotypic limits of the parents.

Translocation the transfer of DNA between nonhomologous chromosomes.

Trihybrid a parental combination that will result in a 27:9:9:9:3:3:3:1 progeny ratio or some variation thereof. Two different alleles are present at each of three independently assorted loci.

Triplex an autopolyploid in which the dominant allele at a locus is present three times.

True breeding progeny will have exactly the same genotype as the parents.

Univalent a single unpaired chromosome during meiosis.

Variance deviation from an average value.

Variety a reproducible plant population for commercial production.

Virulence the ability of a disease organism to attack a host plant.

Visual selection visual identification of desirable genotypes.

Wide crosses combinations between individuals generally at the interspecific or intergeneric levels.

"X" identification for numbers of genomes.

Zygote the initial product of gamete combination.

Index

WIDENER COLLEGE
WOLFGRAM
LIBRARY
CHESTER, PA.

DATE DUE